Undergraduate Lecture Notes in Physics

For further volumes:
http://www.springer.com/series/8917

Undergraduate Lecture Notes in Physics (ULNP) publishes authoritative texts covering topics throughout pure and applied physics. Each title in the series is suitable as a basis for undergraduate instruction, typically containing practice problems, worked examples, chapter summaries, and suggestions for further reading.

ULNP titles must provide at least one of the following:

- An exceptionally clear and concise treatment of a standard undergraduate subject.
- A solid undergraduate-level introduction to a graduate, advanced, or non-standard subject.
- A novel perspective or an unusual approach to teaching a subject.

ULNP especially encourages new, original, and idiosyncratic approaches to physics teaching at the undergraduate level.

The purpose of ULNP is to provide intriguing, absorbing books that will continue to be the reader's preferred reference throughout their academic career.

Series Editors

Neil Ashby
Professor, Professor Emeritus, University of Colorado Boulder, CO, USA

William Brantley
Professor, Furman University, Greenville, SC, USA

Michael Fowler
Professor, University of Virginia, Charlottesville, VA, USA

Michael Inglis
Associate Professor, SUNY Suffolk County Community College, Selden, NY, USA

Elena Sassi
Professor, University of Naples Federico II, Naples, Italy

Helmy Sherif
Professor Emeritus, University of Alberta, Edmonton, AB, Canada

Robert L. Brooks

The Fundamentals of Atomic and Molecular Physics

 Springer

Robert L. Brooks
Department of Physics
University of Guelph
Guelph, Ontario, Canada

ISSN 2192-4791   ISSN 2192-4805 (electronic)
ISBN 978-1-4614-6677-2   ISBN 978-1-4614-6678-9 (eBook)
DOI 10.1007/978-1-4614-6678-9
Springer New York Heidelberg Dordrecht London

Library of Congress Control Number: 2013931254

© Springer Science+Business Media New York 2013
This work is subject to copyright. All rights are reserved by the Publisher, whether the whole or part of the material is concerned, specifically the rights of translation, reprinting, reuse of illustrations, recitation, broadcasting, reproduction on microfilms or in any other physical way, and transmission or information storage and retrieval, electronic adaptation, computer software, or by similar or dissimilar methodology now known or hereafter developed. Exempted from this legal reservation are brief excerpts in connection with reviews or scholarly analysis or material supplied specifically for the purpose of being entered and executed on a computer system, for exclusive use by the purchaser of the work. Duplication of this publication or parts thereof is permitted only under the provisions of the Copyright Law of the Publisher's location, in its current version, and permission for use must always be obtained from Springer. Permissions for use may be obtained through RightsLink at the Copyright Clearance Center. Violations are liable to prosecution under the respective Copyright Law.
The use of general descriptive names, registered names, trademarks, service marks, etc. in this publication does not imply, even in the absence of a specific statement, that such names are exempt from the relevant protective laws and regulations and therefore free for general use.
While the advice and information in this book are believed to be true and accurate at the date of publication, neither the authors nor the editors nor the publisher can accept any legal responsibility for any errors or omissions that may be made. The publisher makes no warranty, express or implied, with respect to the material contained herein.

Printed on acid-free paper

Springer is part of Springer Science+Business Media (www.springer.com)

Preface

This book represents an expansion of a set of course notes for a fourth year undergraduate course in atomic and molecular physics. It assumes two semesters of quantum mechanics as background and could just as easily be called an applied quantum mechanics text. It presents material central to an understanding of structure for both atoms and molecules, developed with a thoroughness not seen in texts since the classics of John C. Slater. It makes no attempt to cover scattering or the multitude of modern topics related to trapping, cooling, or condensation. When used for a 12 week course at the senior undergraduate level, a term paper on some modern topic of the student's interest has been assigned as a supplement and together offer an excellent grounding for students interested in graduate work, whether in this area or some other. Indeed, most of the students taking this course have gone on to study other areas of physics.

The quantum mechanics of complex atoms is not easy to grasp when only cursory or simplified explanations are offered. There seems to be some tacit assumption among authors that only quantum chemists need to know this material and so it is given short shrift in most texts when treated at all. The frustrating thing for many students is that graduate work often assumes that they know this material and yet it is developed from the basics in no book at this level. Whereas many texts develop the two-electron atom using techniques that are not applicable to the many-electron atom, this one treats the two-electron atom as the simplest example of the multi-electron atom and then turns to carbon, as an example, without needing to develop additional equations.

Perturbation techniques are then used to treat fine-structure, the Zeeman and Stark effects, and hyperfine structure. Complications that arise from intermediate coupling or from external fields are handled by direct diagonalization and, for fine structure, are then compared with the results from first-order perturbation.

Spontaneous emission from an atom or molecule in an excited state is another fundamental process which is not often developed from the foundations of time-dependent perturbation theory through the expression for the lifetime of the excited state. Advanced texts can start with Fermi's golden rule while quantum mechanics texts often end there. Developing these expressions in detail is good pedagogy for the student. Asking the question, why does an excited atom decay at all, can stimulate the student to learn quantum electrodynamics even as the answer can be understood, though incompletely, without that.

The electronic structure of diatomic molecules is not so easily accessible as the ro-vibrational interactions and so most books will start with the latter. But if one has just completed a study of the electronic structure of atoms, then to start with the electronic structure of the simplest molecule, H_2^+, makes a lot of sense. Prolate spheroidal coordinates are used, which are natural to the problem, and afford the student usually the first example of performing quantum mechanics using coordinates other than Cartesian, cylindrical, or spherical. The student can perform all of the needed integrals. After that the H_2 molecule is taken up which becomes the molecular analogue of progressing from the hydrogen atom (one electron system) to the complex atom (multi-electron system). One cannot overstate the usefulness, toward understanding molecular bonding, of solving the quantum mechanics of the hydrogen molecular ion and molecule. The ro-vibrational excitations of diatomic molecules are taken up in the final chapter in sufficient detail to satisfy the needs of those progressing toward further study as well as for those not likely to see this material in graduate school.

For most of the years that this material has been used for a fourth year elective course at the University of Guelph it has attracted between 10 and 20 students with the latest numbers nearing thirty. The overwhelming majority of students have gone on to other areas of physics and many have returned to say that this course was where they learned quantum mechanics. I can think of no higher praise.

Guelph, ON, Canada   Robert L. Brooks

Acknowledgements

Turning my lecture notes into a textbook started when one of my students, Theo Hopman, took it upon himself to type up my handwritten notes in LaTeX format. But the first serious suggestion and real impetus for this project was made by Gordon Drake and I am grateful to both of them. Dennis Tokaryk has used the original notes for a number of years and provided invaluable feedback. More than a nod of appreciation should go to my editors Jace Harker and HoYing Fan at Springer. Finally, thanks to the dozens of students who have pointed out typos, mistakes, and confusions whose resolution has greatly improved the final text.

Contents

Part I
Atoms

Chapter 1
Central Forces and Angular Momentum

Much of the material in Chap. 1 will be a review for many of the students using this text. However, indicial notation for vectors will be used throughout this chapter and much of the book, and while most students have been exposed to this notation previously, experience has shown that many have yet to master it. It is not at all difficult and allows for such straightforward proofs of angular momentum relations that it is highly effective to become proficient with it.

Once the commutator relations have been defined, the orbital angular momentum is introduced along with the auxiliary raising and lowering operators. This material is then applied to the solution of the hydrogen atom before generalizing angular momentum through the introduction of spin. Angular momentum is then concluded by considering the addition of two general angular momenta, the definitions and relations regarding Clebsch–Gordan coefficients and the Wigner–Eckart theorem. Hydrogen is then revisited to examine the consequences of spin on its solution, and the basis for the multi-electron treatment of atoms will have been established.

1.1 Indicial Notation for Vectors

A vector $\vec{A}$ has Cartesian coordinates

$$A_x \equiv A_1$$

$$A_y \equiv A_2$$

$$A_z \equiv A_3$$

$\vec{A}$ may be written A_i where i can take any of the values 1, 2, or 3.[1] The inner product of two vectors may be written

---

[1]This ignores the distinction between covariant and contravariant basis vectors which is valid and commonly done when working in three dimensions with orthogonal unit vectors.

R.L. Brooks, *The Fundamentals of Atomic and Molecular Physics*, Undergraduate Lecture Notes in Physics, DOI 10.1007/978-1-4614-6678-9_1, © Springer Science+Business Media New York 2013

$$\vec{A} \cdot \vec{B} = \sum_i A_i B_i \equiv A_i B_i$$

This is the Einstein summation convention. *Repeated Roman subscripts* are summed over. If one wanted to talk about the product of any two like elements, such as $A_2 B_2$ for 1, 2, or 3, one writes $A_\alpha B_\alpha$; i.e., Greek subscripts are not summed. Furthermore if one wanted to refer to a single component of a vector, a Greek rather than a Roman subscript would be used. Also

$$\vec{A} \cdot \vec{B} = \delta_{ij} A_i B_j = A_i B_i \quad \text{or} \quad A_j B_j.$$

Repeated indices are said to be *dummy*; any letter will do. The product is a scalar, is no longer a vector, and is said to be contracted:

$$\delta_{ij} \equiv 1 \quad \text{when} \quad i = j$$

$$0 \quad \text{when} \quad i \neq j$$

The cross product of two vectors is given by

$$\vec{A} \times \vec{B} = \epsilon_{ijk} A_j B_k \quad \text{(summation implied)}$$

where ϵ_{ijk}, the alternating unit tensor, is defined by

$$\epsilon_{ijk} = 0 \quad \text{if any two indices are the same}$$

$$= +1 \quad \text{for} \quad \epsilon_{123}, \epsilon_{231}, \epsilon_{312} \quad \text{cyclic permutation}$$

$$= -1 \quad \text{for} \quad \epsilon_{213}, \epsilon_{321}, \epsilon_{132} \quad \text{anticyclic permutation}$$

A particularly important relationship that will be used for much of the manipulation that follows is

$$\epsilon_{ijk} \epsilon_{ilm} = \delta_{jl} \delta_{km} - \delta_{jm} \delta_{kl}$$

This relationship is tedious but not difficult to prove. Break it down into cases and use Greek indices, for which summation does not apply. For example, case 1 could be that the second two indices of one ϵ are equal: $\epsilon_{ijk} \epsilon_{i\alpha\alpha}$. Clearly this is zero. Check to see what the RHS is. This case covers 3 of the 81 equations represented by the given expression (α can be 1, 2, or 3). But since it doesn't matter which of the two alternating unit tensors has equal second and third indices, this case covers nine equations. In this manner the expression may be proved.

To see the power of this notation, consider the vector identity

$$\vec{\nabla} \cdot (\vec{\nabla} \times \vec{A}) = 0$$

In indicial notation this may be written

$$\nabla_i \epsilon_{ijk} \nabla_j A_k = \epsilon_{ijk} \nabla_i \nabla_j A_k$$

Because the expression is a scalar, all indices are dummy. Change i to j everywhere and permute the gradients:

$$\epsilon_{jik}\nabla_j\nabla_i A_k = \epsilon_{jik}\nabla_i\nabla_j A_k$$

Now when j and i are exchanged in epsilon, the sign changes, yielding the negative of the original expression:

$$\epsilon_{jik}\nabla_i\nabla_j A_k = -\epsilon_{ijk}\nabla_i\nabla_j A_k = 0.$$

1.2 Commutator Algebra

Two operators $\mathbf{A}$ and $\mathbf{B}$ have a commutator defined as

$$[\mathbf{A}, \mathbf{B}] = \mathbf{AB} - \mathbf{BA}$$

It follows that

$$[\mathbf{A}, \mathbf{B}] = -[\mathbf{B}, \mathbf{A}] \tag{1.1}$$

$$[\mathbf{A}, \mathbf{B} + \mathbf{C}] = [\mathbf{A}, \mathbf{B}] + [\mathbf{A}, \mathbf{C}] \tag{1.2}$$

$$[\mathbf{A}, \mathbf{BC}] = [\mathbf{A}, \mathbf{B}]\mathbf{C} + \mathbf{B}[\mathbf{A}, \mathbf{C}] \tag{1.3}$$

Equation (1.3) can be readily proven:

$$
\begin{aligned}
[\mathbf{A}, \mathbf{BC}] &= \mathbf{ABC} - \mathbf{BCA} \\
&= \mathbf{ABC} - \mathbf{BAC} + \mathbf{BAC} - \mathbf{BCA} \\
&= (\mathbf{AB} - \mathbf{BA})\mathbf{C} + \mathbf{B}(\mathbf{AC} - \mathbf{CA}) \\
&= [\mathbf{A}, \mathbf{B}]\mathbf{C} + \mathbf{B}[\mathbf{A}, \mathbf{C}]
\end{aligned}
$$

Consider the position operator q_i where $q_1 \equiv x$, $q_2 \equiv y$, $q_3 \equiv z$ and the momentum operator p_i where

$$p_1 = -i\hbar\frac{\partial}{\partial x}, \quad p_2 = -i\hbar\frac{\partial}{\partial y}, \quad p_3 = -i\hbar\frac{\partial}{\partial z}$$

$$\text{Then} \quad p_i \equiv -i\hbar\frac{\partial}{\partial q_i} \quad \text{or} \quad -i\hbar\nabla_i$$

$$\text{and} \quad [q_i, q_j] = 0 \tag{1.4}$$

$$[p_i, p_j] = 0 \tag{1.5}$$

$$[q_i, p_j] = i\hbar\delta_{ij}. \tag{1.6}$$

To prove this last relation, invoke some arbitrary function A which depends on variables q_i. A is introduced *only* for pedagogical reasons:

$$[q_i, p_j]\, A = -i\hbar q_i \frac{\partial}{\partial q_j} A + i\hbar \frac{\partial}{\partial q_j} q_i A$$

The second term is $\quad i\hbar q_i \dfrac{\partial}{\partial q_j} A + i\hbar \left(\dfrac{\partial}{\partial q_j} q_i \right) A$

So $\quad [q_i, p_j]\, A = i\hbar \left(\dfrac{\partial}{\partial q_j} q_i \right) A = i\hbar\delta_{ij} A$

$$[q_i, p_j] = i\hbar\delta_{ij}$$

$$[q_i, F(q_j)] = 0 \tag{1.7}$$

$$[p_i, F(p_j)] = 0 \tag{1.8}$$

$$[p_i, F(q_j)] = -i\hbar \frac{\partial}{\partial q_i} F(q_j) \tag{1.9}$$

Proof of Eq. (1.9) is done just like (1.6).

1.3   Orbital Angular Momentum

Angular momentum in quantum mechanics plays a most important role. If one starts with the classical expression and makes the standard conversion for the momentum operator, one obtains a quantum mechanical operator for orbital angular momentum. What is important to keep in mind is that angular momentum in quantum mechanics is a more general concept than just this. How that comes about will be presented as this chapter proceeds.

Consider the quantum Hamiltonian of a particle with mass m in the central potential $V(r)$:

$$\mathbf{H} = \frac{p^2}{2m} + V(r) = -\hbar^2 \frac{\nabla^2}{2m} + V(r)$$

The problem of an electron bound to a proton can be described by such a Hamiltonian when $V(r)$ is a Coulomb potential. Chapter 2 will consider this problem for more than one electron. Remembering that the axioms of quantum mechanics permit simultaneous determinations of measurable quantities only if those quantities are observables described by commuting operators, it becomes important to investigate the operators that commute with the Hamiltonian and with each other. The orbital angular momentum, classically conserved for the central force problem (Kepler's Second Law), is

$$\vec{L} = \vec{r} \times \vec{p} = \vec{r} \times \frac{\hbar}{i} \vec{\nabla}$$

$$\text{or} \quad L_i = \epsilon_{ijk} q_j p_k$$

What are the commutation relations for $\vec{L}$?

$$[L_i, q_l] = \epsilon_{ijk} [q_j p_k, q_l]$$

$$= \epsilon_{ijk} q_j [p_k, q_l] + \epsilon_{ijk} \underbrace{[q_j, q_l]}_{=0} p_k$$

$$= -i\hbar \epsilon_{ijk} q_j \delta_{kl}$$

$$= i\hbar \epsilon_{ilj} q_j$$

Changing free and dummy indices gives

$$[L_i, q_j] = i\hbar \epsilon_{ijk} q_k \tag{1.10}$$

This illustrates a typical situation that occurs when using indicial notation. It also affords an opportunity to emphasize an important rule relating to the use of these indices. The two expressions, before and after changing the indices, which I shall write again here, are identical:

$$[L_i, q_l] = i\hbar \epsilon_{ilj} q_j \qquad [L_i, q_j] = i\hbar \epsilon_{ijk} q_k$$

First realize that one *never* changes indices in the middle of a derivation. One starts a derivation with some letters assigned to indices because they were the ones used in a definition (like j and k above). Those letters, being dummy, may drop out over the course of the derivation of the new expression, as k did above. You would like to express your answer using letters that are familiar, in a familiar order, like i, j, k. To do that you reexpress the dummy index j as k because that letter is available and then change the free index l to j everywhere it appears because now that letter too is available. It is important to realize that such changes are for appearance only, usually just to check your answer, and have no further significance.

Look again at the two expressions above. For the first one, i and l are free indices because they appear once on each side of the equal sign. j is a dummy index because it appears twice in a product. Any index that appears three or more times in a product is an error. No index may appear more than twice. So the dummy index j could be changed to any letter, so change it to k. Our entity, which is a commutator of two vectors having indices i and l, has not been effected. But why should the commutator have those indices? They were freely chosen and could have been any two letters at all. If one chooses to reexpress that commutator using different letters, then one better change those letters everywhere they appear on both sides of the equation. Changing l to j, which is allowed because the letter j is no longer being used, then permits the result to be expressed using the subscripts i, j, and k.

Problem 1.1

Show that

$$[L_i, p_j] = i\hbar\epsilon_{ijk}p_k \tag{1.11}$$

$$[L_i, L_j] = i\hbar\epsilon_{ijk}L_k. \tag{1.12}$$

Remember that this means $[\mathbf{L}_x, \mathbf{L}_y] = i\hbar\mathbf{L}_z$, for example.

Problem 1.2

Use the second relation above to show that $\vec{\mathbf{L}} \times \vec{\mathbf{L}} = i\hbar\vec{\mathbf{L}}$.

So the components of angular momentum do *not* commute with the position, the linear momentum, or with each other. However,

$$[L_i, p_j^2] = [L_i, p_j]\, p_j + p_j\,[L_i, p_j]$$

$$= 2i\hbar\epsilon_{ijk}p_k p_j = 0.$$

Why? It is because of the sum over j and k. The only nonzero values are

$$\epsilon_{\alpha\beta\gamma}p_\beta p_\gamma + \epsilon_{\alpha\gamma\beta}p_\beta p_\gamma = \epsilon_{\alpha\beta\gamma}p_\beta p_\gamma - \epsilon_{\alpha\beta\gamma}p_\beta p_\gamma = 0.$$

The components of angular momentum *do* commute with the kinetic energy. Similarly, $[L_i, q_j^2] = 0$.

What about $[L_i, r] = \left[L_i, (q_j^2)^{1/2}\right]$?

$$[L_i, r] = \epsilon_{ijk}q_j\left[p_k, (q_l^2)^{1/2}\right]$$

$$= -i\hbar\epsilon_{ijk}q_j\nabla_k(q_l^2)^{1/2} \quad \text{By property (1.9)}$$

$$= -i\hbar\epsilon_{ijk}q_j\frac{q_l}{r}\delta_{lk}$$

$$= -i\hbar\epsilon_{ijk}\frac{q_j q_k}{r} = 0$$

So $\vec{\mathbf{L}}$ commutes with r. In addition, $\vec{\mathbf{L}}$ commutes with any function of r:

$$[L_i, V(r)] = -i\hbar\epsilon_{ijk}q_j\nabla_k V(r)$$

$$\text{But} \quad \frac{\partial}{\partial q_k}V(r) = \frac{\partial V}{\partial r}\frac{\partial r}{\partial q_k} = \frac{q_k}{r}\frac{\partial}{\partial r}V(r)$$

$$\text{So} \quad [L_i, V(r)] = -i\hbar\epsilon_{ijk}\frac{q_j q_k}{r}\frac{\partial}{\partial r}V(r) = 0$$

One consequence of $\vec{\mathbf{L}}$ commuting with any function of r is that each component of $\vec{\mathbf{L}}$ commutes with the Hamiltonian for central force problems:

$$[L_i, \mathbf{H}] = 0 \quad \text{for central forces.}$$

With what else does $\vec{\mathbf{L}}$ commute? Define $\mathbf{L}^2 = \mathbf{L}_x^2 + \mathbf{L}_y^2 + \mathbf{L}_z^2 = \mathbf{L}_i^2$:

$$\begin{aligned}
\left[\mathbf{L}^2, \vec{\mathbf{L}}\right] &= \left[L_i^2, L_j\right] \\
&= L_i[L_i, L_j] + [L_i, L_j]L_i \\
&= i\hbar\epsilon_{ijk}L_i L_k + i\hbar\epsilon_{ijk}L_k L_i \\
&= i\hbar\epsilon_{jki}(L_i L_k + L_k L_i) \\
&= 0
\end{aligned}$$

Be careful: unlike previous derivations, L_i and L_k do *not* commute. The pedestrian way to show that the above commutator in fact is 0 is to consider $j = 1$; this then gives

$$j = 1: \quad i\hbar(L_3 L_2 - L_2 L_3 + L_2 L_3 - L_3 L_2) = 0$$

Similarly for $j = 2$ and $j = 3$, so $\left[\mathbf{L}^2, \vec{\mathbf{L}}\right] = 0$. One could also note that because k and i are dummy, interchanging them everywhere and then interchanging them just in the alternating unit tensor gives

$$-i\hbar\epsilon_{jki}[L_k L_i + L_i L_k].$$

But this is the negative of the above expression (the order of the double operator expressions in the sum is unimportant), and hence it equals 0. Also, $\mathbf{L}^2$ commutes with each component of angular momentum. Since $\vec{\mathbf{L}}$ commutes with $\mathbf{H}$, $\mathbf{L}^2$ must also commute with $\mathbf{H}$. (Why?)

Now let's look at the explicit form for the angular momentum operators in spherical coordinates:

$$\vec{\nabla} = \hat{r}\frac{\partial}{\partial r} + \hat{\phi}\frac{1}{r\sin\theta}\frac{\partial}{\partial\phi} + \hat{\theta}\frac{1}{r}\frac{\partial}{\partial\theta}$$

$$\hat{r} = \sin\theta\cos\phi\hat{x} + \sin\theta\sin\phi\hat{y} + \cos\theta\hat{z}$$

$$\hat{\theta} = \cos\theta\cos\phi\hat{x} + \cos\theta\sin\phi\hat{y} - \sin\theta\hat{z}$$

$$\hat{\phi} = -\sin\phi\hat{x} + \cos\phi\hat{y}$$

$$\text{Then}\quad \vec{L} = \vec{r}\times\vec{p} = r\hat{r}\times\frac{\hbar}{i}\vec{\nabla}$$

$$= \frac{\hbar}{i}\left(\hat{\phi}\frac{\partial}{\partial\theta} - \hat{\theta}\frac{1}{\sin\theta}\frac{\partial}{\partial\phi}\right)$$

In this form it is apparent that $\vec{L}$ commutes with any function of r.

From these expressions it is easy to find $\mathbf{L}_x$, $\mathbf{L}_y$, and $\mathbf{L}_z$ and is also straightforward to show

$$\mathbf{L}^2 = -\hbar^2\left[\frac{1}{\sin^2\theta}\frac{\partial}{\partial\phi^2} + \frac{1}{\sin\theta}\frac{\partial}{\partial\theta}\left(\sin\theta\frac{\partial}{\partial\theta}\right)\right].$$

The eigenvalue problem for the angular momentum operators $\mathbf{L}^2$ and $\mathbf{L}_z$ is fundamental to all central force problems. Since $\mathbf{L}^2$ and $\mathbf{L}_z$ commute, it is possible to find an eigenfunction that is simultaneously a solution for each operator. The spherical harmonics form such a solution:

$$\boxed{\begin{array}{l} \mathbf{L}^2 \mathrm{Y}_\ell^m(\theta,\phi) = \ell\,(\ell+1)\,\hbar^2 \mathrm{Y}_\ell^m(\theta,\phi) \\[2mm] \mathbf{L}_z \mathrm{Y}_\ell^m(\theta,\phi) = m\hbar \mathrm{Y}_\ell^m(\theta,\phi) \end{array}}$$

$$\text{(1.13)}$$
$$\text{(1.14)}$$

Here the symbol ℓ is referred to as the orbital angular momentum even though it is not itself an eigenvalue of any operator. $\ell\,(\ell+1)$ is the eigenvalue of $\mathbf{L}^2$ rather than ℓ^2 as a consequence of the spherical harmonics. Usually textbooks simply say it is a matter of convenience. It certainly is that! m is the eigenvalue of $\mathbf{L}_z$, usually called the magnetic quantum number but that is an unfortunate misnomer. It is better to think of it as a directional quantum number.

Spherical harmonics are simply one member of a family of special functions that form the solutions of the Sturm–Liouville problem. Such functions are the subject of courses in *Special Functions*, and all too often students who have never taken such a course feel intimidated by their use. This is unfortunate because *using* special functions is very different from deriving the relations that they satisfy. Learning to manipulate the relations is not so difficult and is all one needs to understand their use in quantum mechanics. The relations that you need are given in Appendix B.

One of the important concepts in QM is that the eigenfunctions of operators of interest span a Hilbert space. A complete linear vector space with a defined inner product is a Hilbert space. The above equations can be rewritten as

$$\mathbf{L}^2|\ell m\rangle = \ell(\ell+1)\hbar^2|\ell m\rangle \tag{1.15}$$

$$\text{and}\quad \mathbf{L}_z|\ell m\rangle = m\hbar|\ell m\rangle \tag{1.16}$$

The ket $|\ell m\rangle$ is a vector in an infinite dimensional Hilbert space. The commutator relations among the operators $\vec{\mathbf{L}}$ and $\mathbf{L}^2$, along with the relations (1.15) and (1.16), completely define the concept of angular momentum. The functions $Y_\ell^m(\theta,\phi)$ are *one* representation of the kets $|\ell m\rangle$. But the kets have an existence in their Hilbert space that is independent of whether a coordinate representation exists. Angular momentum in QM has a well-defined existence independent of the coordinates r, θ, ϕ (or any other spatial coordinates).

Let us pursue these concepts by examining the ladder operators $\mathbf{L}_+$ and $\mathbf{L}_-$. The auxiliary operators $\mathbf{L}_+$ and $\mathbf{L}_-$ are defined as

$$\mathbf{L}_+ \equiv \mathbf{L}_x + i\mathbf{L}_y$$
$$\mathbf{L}_- \equiv \mathbf{L}_x - i\mathbf{L}_y$$

It is easy to prove the following commutators:

Problem 1.3

$$[\mathbf{L}_z, \mathbf{L}_+] = \hbar\mathbf{L}_+ \tag{1.17}$$

$$[\mathbf{L}_z, \mathbf{L}_-] = -\hbar\mathbf{L}_- \tag{1.18}$$

$$[\mathbf{L}_+, \mathbf{L}_-] = 2\hbar\mathbf{L}_z \tag{1.19}$$

$$[\mathbf{L}^2, \mathbf{L}_\pm] = 0 \tag{1.20}$$

Next take note of the following product of operators:

$$\mathbf{L}_-\mathbf{L}_+ = (\mathbf{L}_x - i\mathbf{L}_y)(\mathbf{L}_x + i\mathbf{L}_y)$$
$$= \mathbf{L}_x^2 + \mathbf{L}_y^2 + i(\mathbf{L}_x\mathbf{L}_y - \mathbf{L}_y\mathbf{L}_x)$$
$$= \mathbf{L}_x^2 + \mathbf{L}_y^2 + i[\mathbf{L}_x, \mathbf{L}_y]$$
$$= \mathbf{L}_x^2 + \mathbf{L}_y^2 - \hbar\mathbf{L}_z$$
$$\mathbf{L}_-\mathbf{L}_+ = \mathbf{L}^2 - \mathbf{L}_z(\mathbf{L}_z + \hbar)$$

Similarly $\mathbf{L}_+\mathbf{L}_- = \mathbf{L}^2 - \mathbf{L}_z(\mathbf{L}_z - \hbar)$. Since these operator products are diagonal using the basis $|\ell m\rangle$, it is straightforward to evaluate the eigenvalues for each:

$$\mathbf{L}_-\mathbf{L}_+|\ell m\rangle = \mathbf{L}^2|\ell m\rangle - \mathbf{L}_z(\mathbf{L}_z + \hbar)|\ell m\rangle$$

$$= \hbar^2\left[\ell(\ell+1) - m(m+1)\right]|\ell m\rangle$$

or $\quad \mathbf{L}_-\mathbf{L}_+|\ell m\rangle = \hbar^2(\ell - m)(\ell + m + 1)|\ell m\rangle$ $\qquad(1.21)$

$$\mathbf{L}_+\mathbf{L}_-|\ell m\rangle = \hbar^2(\ell + m)(\ell - m + 1)|\ell m\rangle \qquad(1.22)$$

Enough has now been developed to prove an important relationship. The eigenvalue m, which can be positive or negative, cannot exceed the value of ℓ; i.e.,

$$-\ell \leq m \leq \ell$$

The proof is straightforward. The norm of the vectors $\mathbf{L}_\pm|\ell m\rangle$ is positive definite or

$$|(\mathbf{L}_\pm|\ell m\rangle)| \geq 0$$

But this can be written as

$$\langle \ell m|\mathbf{L}_\mp\mathbf{L}_\pm|\ell m\rangle \geq 0$$

$$\langle \ell m|\mathbf{L}^2 - \mathbf{L}_z(\mathbf{L}_z \pm \hbar)|\ell m\rangle \geq 0$$

$$\hbar^2\left[\ell(\ell+1) - m(m+1)\right] \geq 0$$

From which it follow that

$$\ell(\ell+1) \geq m^2 + m$$

$$\ell(\ell+1) \geq m^2 - m$$

But there is no reason why m can't be positive, negative, or zero and also must be integer from which the result follows.

Next the relations that the operators $\mathbf{L}_\pm$ themselves satisfy need to be developed. Consider

$$\mathbf{L}_z\mathbf{L}_+|\ell m\rangle = (\hbar\mathbf{L}_+ + \mathbf{L}_+\mathbf{L}_z)|\ell m\rangle \quad \text{from (1.17)}$$

$$= (m+1)\hbar\mathbf{L}_+|\ell m\rangle.$$

But $\mathbf{L}_z|\ell m+1\rangle = (m+1)\hbar|\ell m+1\rangle$. So the vector $\mathbf{L}_+|\ell m\rangle$ must be a scalar multiple of $|\ell m+1\rangle$. Or

$$\mathbf{L}_+|\ell m\rangle = \alpha|\ell m+1\rangle \qquad(1.23)$$

$\mathbf{L}_+$ has raised the m index by one. $\mathbf{L}_+$ is also called a raising operator. Now what is α? The value of α can be found without appeal to any properties of Y_ℓ^m's. Simply take the norm of both sides of Eq. (1.23). Something similar was done above. The rule when taking the complex conjugate using bra–ket notation is to replace every ket by its bra, every number by its complex conjugate, and every operator by its adjoint specified by the dagger. For an operator written in functional notation that simply means replacing it with its complex conjugate. But in bra–ket notation one should think of the operators as matrices in which case the adjoint is the transpose of the complex conjugate:

$$\text{RHS:}\quad \text{Norm} = \langle \ell\, m + 1\, |\alpha^*\alpha|\, \ell\, m + 1 \rangle = \alpha^*\alpha = \alpha^2$$

$$\text{LHS:}\quad \text{Norm} = \langle \ell\, m\, |\mathbf{L}_+^\dagger \mathbf{L}_+|\, \ell\, m \rangle = \langle \ell\, m\, |\mathbf{L}_- \mathbf{L}_+|\, \ell\, m \rangle$$

$$= \hbar^2 (\ell - m)(\ell + m + 1)$$

The phase of α can freely be chosen to be real so that $\alpha = \hbar\sqrt{(\ell - m)(\ell + m + 1)}$. Then

$$\boxed{\mathbf{L}_+|\,\ell\, m \rangle = \hbar\sqrt{(\ell - m)(\ell + m + 1)}|\,\ell\, m + 1 \rangle} \qquad (1.24)$$

$$\boxed{\mathbf{L}_-|\,\ell\, m \rangle = \hbar\sqrt{(\ell + m)(\ell - m + 1)}|\,\ell\, m - 1 \rangle} \qquad (1.25)$$

Since m is bounded, it follows that

$$\mathbf{L}_+|\,\ell\, \ell \rangle = 0$$
$$\mathbf{L}_-|\,\ell - \ell \rangle = 0.$$

Equations (1.24) and (1.25) are more important than they at first appear. For a fixed value of ℓ, the vectors $|\,\ell\, m \rangle$ span a *finite* dimensional subspace of dimension $2\ell + 1$. By somehow finding *one* vector in this subspace, repeated application of Eq. (1.24) or (1.25) will yield all of the other vectors. $\mathbf{L}_+$ and $\mathbf{L}_-$ cannot project a vector out of this subspace because they commute with $\mathbf{L}^2$.

Another obvious consequence of Eqs. (1.24) and (1.25) is that $|\,\ell\, m \rangle$ are not eigenvectors of the operators $\mathbf{L}_\pm$. There is no reason to expect them to be because $\mathbf{L}_\pm$ do not commute with $\mathbf{L}_z$. Furthermore, $\mathbf{L}_\pm$ are not Hermitian operators.

1.4   Solution of the Hydrogen Atom

It is assumed that the student has encountered the hydrogen atom's solution in a previous quantum mechanics course. What is presented here will be a review with emphasis on those aspects necessary for an understanding of the multi-electron problem.

For an electron in a Coulomb potential

$$\mathbf{H}u = Eu$$

$$\mathbf{H} = -\frac{\hbar^2}{2\mu}\nabla^2 - \frac{Ze^2}{4\pi\epsilon_o\, r}$$

$$\nabla^2 = \frac{1}{r^2}\frac{\partial}{\partial r}r^2\frac{\partial}{\partial r} - \frac{\mathbf{L}^2}{\hbar^2 r^2}$$

$$u(r,\theta,\phi) = R(r)\Theta(\theta)\Phi(\phi) = R(r)Y_\ell^m(\theta,\phi)$$

Putting this into the above yields the differential equation for the r variable:

$$\left[\frac{1}{r^2}\frac{d}{dr}r^2\frac{d}{dr} + \frac{2\mu}{\hbar^2}\left(E + \frac{Ze^2}{4\pi\epsilon_o\, r}\right) - \frac{\ell(\ell+1)}{r^2}\right]R(r) = 0$$

Here $\mu = \frac{m_e m_N}{m_e + m_N}$ is the reduced mass. Recall that whenever two bodies interact through a central force, the problem is equivalent to one body having the reduced mass acting under the influence of the same force. For this problem the two bodies differ in mass by nearly a factor of 2,000, and the reduced mass is nearly the mass of the electron. Often one thinks of the electron being electrostatically bound to an infinitely massive nucleus. In Chap. 6 the two bodies will be those of the nuclei of diatomic molecules with similar masses, and a derivation of this consequence will be given.

Atomic units (a.u.)will be introduced at this point. The reader is warned that these units are used by no other humans, that their adoption to other branches of physics is not being encouraged, and that it is not possible to uniquely restore dimensions to any result that one obtains. They are profoundly convenient for theoretical atomic physics, are widely used within that community, and do not present any insurmountable challenge to the student:

$$4\pi\epsilon_o = \hbar = e = m_e = 1$$

$$c \neq 1 \qquad c = \frac{1}{\alpha} \approx 137$$

α is the fine-structure constant:

$$\alpha \equiv \frac{e^2}{4\pi\epsilon_o\hbar c} = \frac{1}{137.0359(3)}$$

The energy unit is the Hartree:

$$1\ \text{Hartree} = 2\ \text{Rydbergs} = \frac{m_e e^4}{(4\pi\epsilon_o)^2\hbar^2} = 27.21140\ \text{eV}$$

$$a_{o} = \text{Bohr radius} = \frac{4\pi\epsilon_{o}\hbar^2}{m_e e^2} = 1 \quad \text{length unit}$$

$$= 0.529178\,\text{Å}$$

$$= 5.29178 \cdot 10^{-11}\,\text{m}$$

Problem 1.4

Show that one Hartree can be written as $\alpha^2 m_e c^2$.

A simple fact of life for a research physicist is that one needs to be proficient using different systems of units, particularly energy units. The Hartree is the unit used for all atomic, theoretical calculations. The electron volt (eV) is the unit most widely used in other branches of physics, and the cm^{-1} (to be introduced subsequently) is favored by spectroscopists.

The differential equation (DE) for $R(r)$ may be written for a nucleus of infinite mass in atomic units as

$$\frac{1}{r^2}\frac{d}{dr}\left(r^2\frac{d}{dr}\right) + 2\left[E + \frac{Z}{r} - \frac{\ell(\ell+1)}{2r^2}\right]R = 0$$

Introducing $P(r)$ here allows a somewhat simpler DE to be obtained. Please be alert to the fact that both $R(r)$ and $P(r)$ are referred to as hydrogenic radial wave functions. Tables are as likely to list the one as the other:

$$\text{Let} \quad P(r) \equiv rR(r)$$

$$\frac{dP}{dr} = r\frac{dR}{dr} + R$$

$$\frac{d^2P}{dr^2} = r\frac{d^2R}{dr^2} + 2\frac{dR}{dr} = \frac{1}{r}\frac{d}{dr}\left(r^2\frac{dR}{dr}\right)$$

So the above DE becomes

$$\frac{d^2P}{dr^2} + 2\left[E + \frac{Z}{r} - \frac{\ell(\ell+1)}{2r^2}\right]P = 0$$

Conditions placed on the solution (finite at ∞, continuous, etc.) demand that the series which expresses the solution should terminate. This termination introduces another integer quantity n, which is needed (besides ℓ) to define P. These conditions also lead to

$$\boxed{E = -\frac{Z^2}{2n^2}\ \text{Hartrees}} \tag{1.26}$$

for an infinite nuclear mass. For a finite nuclear mass the expression is

$$E = -\frac{\mu Z^2}{2m_e n^2} = -\frac{RZ^2}{n^2}.$$

The second version uses R which is the Rydberg constant. This constant, expressed in some energy unit, has different values for different elements, e.g.,

$$R_\infty = 109737.32 \,\text{cm}^{-1}$$

$$R_H = 109677.58 \,\text{cm}^{-1}$$

The energy unit cm^{-1} comes from the Planck relation $E = h\nu = hc/\lambda$.

$\dfrac{1}{\lambda} = \dfrac{E}{hc}$ express λ in cm and the others in compatible units.

This expression can be a source of some confusion, so let me try to clarify. Energy can be expressed in many different units. You should be familiar with joules and electron volts and probably ergs. In atomic units, energy is expressed in Hartrees, is introduced above, and is dimensionless. If one looks at the Planck relation and realizes that both h and c are universal constants, one could convert any energy into either Hz or cm^{-1}. Such a conversion has nothing to do with atomic units and is done simply for laboratory convenience. Because many of the results of atomic physics are used by spectroscopists and they like the unit cm^{-1}, that unit will often (certainly not always) be used in this text when comparing to experiment. Some authors make a distinction between the Rydberg constant expressed in cm^{-1} given above and the same constant expressed in joules or eV which is sometimes written "Ry." Such a distinction will not be used in this text. Figure 1.1 presents a Grotrian level diagram for the energy levels of hydrogen. This type of diagram is common among spectroscopists and is one in which the energy is graphed vertically while distinct values of angular momentum are graphed horizontally. Lines are often drawn between allowed transitions, and the wavelengths of the transitions are written along the line.

The solution for $P(r)$ may be expressed as

$$\text{P}_{n\ell}(r) = \frac{-\left[(n - \ell - 1)!Z\right]^{1/2}}{n\left[(n + \ell)!\right]^{1/2}} x^{\ell+1} e^{-x/2} \text{L}_{n-\ell-1}^{2\ell+1}(x) \tag{1.27}$$

where $x \equiv 2Zr/n$ and $\text{L}(x)$ are the associated Laguerre polynomials. The $\text{P}_{n\ell}$'s have the following orthonormality condition:

$$\int_0^\infty \text{P}_{n\ell}(r)\text{P}_{n'\ell}(r)\, dr = \delta_{nn'} \tag{1.28}$$

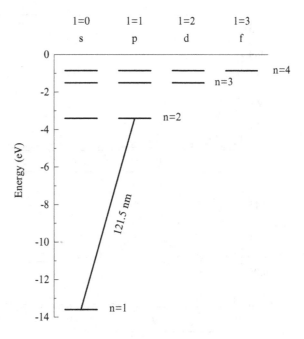

Fig. 1.1 Grotrian level diagram for hydrogen

Recall that

$$\int_0^\pi \sin\theta\, d\theta \int_0^{2\pi} Y_\ell^m(\theta,\phi) Y_{\ell'}^{m'}(\theta,\phi)\, d\phi = \delta_{\ell\ell'}\delta_{mm'} \qquad (1.29)$$

Finally, the eigenfunctions for the hydrogen atom are

$$U(r,\theta,\phi) = \frac{P_{n\ell}(r)}{r} Y_\ell^m(\theta,\phi) \qquad (1.30)$$

The squared, normalized radial wave functions multiplied by r^2, the radial part of the volume element, $P_{n\ell}^2$ are graphed in Fig. 1.2. The top panel shows the s-states, the middle panel the p-states, and the bottom the d-states for a selection of n values. Note that the number of nodes, the zeros of the wave functions, are given by $n-\ell-1$.

It is obviously more difficult to attempt to graph the wave function, or its square, if one wants to include the angular part. In *The Picture Book of Quantum Mechanics*,[2] the authors show a number of different projections of hydrogenic wave functions. The one that graphs r and θ as polar variables in the $x-z$ plane with the probability density perpendicular to this plane is particularly effective since the azimuthal variable is often unimportant because of symmetry. It is interesting to note that the appearance of such a graph is remarkably different depending on

---

[2]Siegmund Brandt and Hans Dieter Dahmen, Springer-Verlag, New York, 1995.

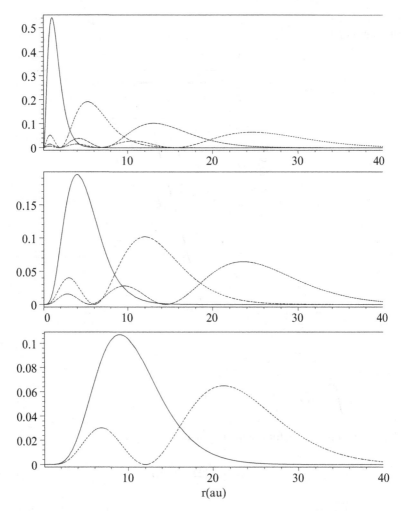

Fig. 1.2 Squared, radial wave functions, $P_{n\ell}^2$, for 1s–4s (*top*), 2p–4p (*middle*), and 3d–4d (*bottom*) for hydrogen

whether one includes or excludes the volume element component $r^2 \sin\theta$. Since that was included in the graphs of the squares of the radial wave functions, the volume element will also be included in the following graphs. As a caution to the reader, this is *not* the convention followed in *The Picture Book*.

Figure 1.3 shows the square of the 3p $m = 1$ hydrogenic wave function multiplied by the volume element. Figure 1.4 is the same but for $m = 0$. One might think there would be no difference since the azimuthal variable is suppressed in the plots, but a glance at Table 1.3 clearly shows that Y_1^1 differs in its functional form for θ from Y_1^0 and this causes the observed difference between these plots. Figure 1.5

Fig. 1.3 The $n = 3$, $\ell = 1$, $m_\ell = 1$ hydrogenic probability density

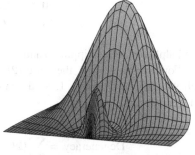

Fig. 1.4 The $n = 3$, $\ell = 1$, $m_\ell = 0$ hydrogenic probability density

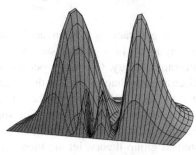

Fig. 1.5 The $n = 3$, $\ell = 0$, $m_\ell = 0$ hydrogenic probability density

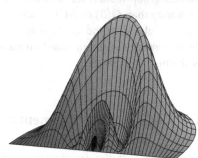

shows the square of the 3s hydrogenic wave function, and the three bulges that show so prominently in Fig. 1.2 can also be seen here.

An eigenstate of hydrogen may be labeled

$$\mathbf{H}|\,n\,\ell\,m\,\rangle = E|\,n\,\ell\,m\,\rangle$$

For a given n, $\ell \leq n - 1$. So for

$$n = 1\; \ell = 0 \quad m = 0$$
$$n = 2\; \ell = 0, 1\; m = -1, 0, 1$$

etc.

Spectroscopic notation for ℓ:

$$\ell = \begin{array}{ccccccccc} 0 & 1 & 2 & 3 & 4 & 5 & 6 & 7 & 8 \\ s & p & d & f & g & h & i & k & l \end{array}$$

and alphabetically skipping p and s.

For each value of ℓ, there are $(2\ell + 1)$ values for m. For a given n, ℓ takes on all values from 0 to $n - 1$. So the number of states sharing a common energy, the degeneracy, is

$$\text{Degeneracy} = \sum_{\ell=0}^{n-1} (2\ell + 1) = 2\left[\frac{(n-1)n}{2}\right] + n = n^2$$

This degeneracy is *not* an accidental degeneracy. By that is meant that two energy levels in a given system with different quantum numbers happen to have the same energy. Often this occurs when applying external fields to an atom as discussed in Chap. 3. Rather, this degeneracy is essential arising from a symmetry that the Coulomb field has. There is a fundamental connection between essential degeneracies and symmetries. The study of group theory, particularly continuous groups, explores that connection. While it is not expected that the student has studied group theory, let me mention that the symmetry group is $O(4)$, a higher symmetry than $O(3)$ or $SU(2)$ which are common to the central field problem and may be related to the isotropy of space. This latter symmetry is responsible for the degeneracy of the m quantum number which occurs even in the multi-electron problem.

1.5   Spin Angular Momentum

One reason for the frequent confusion when spin is introduced is that wave mechanics cannot handle the concept naturally, and so it often appears like a fifth wheel on a wagon. Matrix mechanics is where it belongs and where the appearance is really rather natural.

What if the matrix element for a given operator $\mathbf{A}$ with respect to some known wave function were desired? One could write

$$\langle n \,|\mathbf{A}|\, n' \rangle = \int \Psi_n^*(\vec{r}) \mathbf{A} \Psi_{n'}(\vec{r}) \, d\tau$$

By arranging the indices n and n' into rows and columns, we could evaluate each element in turn and fill a table so long as n and n' had finite dimensions.

What if the wave functions were hydrogenic orbitals? Then there would be

$$\langle n\,\ell\,m\,|\mathbf{A}|\,n'\,\ell'\,m'\rangle = \int u^*_{n\ell m}(r,\theta,\phi)\mathbf{A}u_{n'\ell'm'}(r,\theta,\phi)\,d\tau$$

How can this be written as a matrix? Since all of the m and ℓ values are bounded, it could be built up something like

		n'	1	2	2			3
		ℓ'	0	0	1			0
n	ℓ	$m\backslash m'$	0	0	−1	0	+1	0
1	0	0						
2	0	0						
		−1						$\langle 2\,1-1\,|\mathbf{A}|\,3\,0\,0\rangle$
2	1	0						
		+1						
3	0	0						

Clearly the matrix is infinite, cumbersome, and not at all practical. The matrices for particular subspaces can, however, be useful. Consider the angular momentum portion of the Hilbert space, and ask what is the matrix of $\langle \ell\,m\,|\mathbf{L}_x|\,\ell\,m'\rangle$ for, say, $\ell = 1$. (Since $\mathbf{L}_x$ does not operate on the n part, n must equal n'.)

The matrix under investigation is

ℓ		1	
$\ell\;m\backslash m'$	−1	0	+1
−1	$\langle 1-1\,|\mathbf{L}_x|\,1-1\rangle$	$\langle 1-1\,|\mathbf{L}_x|\,1\,0\rangle$	$\langle 1-1\,|\mathbf{L}_x|\,1\,1\rangle$
1 0	$\langle 1\,0\,|\mathbf{L}_x|\,1-1\rangle$	$\langle 1\,0\,|\mathbf{L}_x|\,1\,0\rangle$	$\langle 1\,0\,|\mathbf{L}_x|\,1\,1\rangle$
+1	$\langle 1\,1\,|\mathbf{L}_x|\,1-1\rangle$	$\langle 1\,1\,|\mathbf{L}_x|\,1\,0\rangle$	$\langle 1\,1\,|\mathbf{L}_x|\,1\,1\rangle$

Now

$$\langle \ell\,m\,|\mathbf{L}_x|\,\ell\,m'\rangle = \int Y^{*m}_\ell \mathbf{L}_x Y^{m'}_\ell\,d\Omega$$

This equality is purely formal and merely demonstrates the equivalence of matrix mechanics and wave mechanics. These matrix elements could be evaluated either way.

Using wave mechanics,

$$\mathbf{L}_x = i\hbar\left(\sin\phi\frac{\partial}{\partial\theta} + \cot\theta\cos\phi\frac{\partial}{\partial\phi}\right)$$

Consider, e.g.,

$$\int Y_1^{1*} \mathbf{L}_x Y_1^0 \, d\Omega$$

$$= -\left(\frac{3}{8\pi}\right)^{1/2} \int \sin\theta e^{-i\phi} \left[i\hbar \left(\sin\phi \frac{\partial}{\partial\theta} + \cot\theta \cos\phi \frac{\partial}{\partial\phi} \right) \right] \left(\frac{3}{4\pi}\right)^{1/2} \cos\theta \, d\Omega$$

$$= \frac{3i\hbar}{\sqrt{24\pi}} \int \sin^2\theta \sin\phi e^{-i\phi} \, d\Omega$$

$$= \frac{3\hbar}{4\sqrt{2}} \int \sin^3\theta \, d\theta$$

$$= \frac{\hbar}{\sqrt{2}}$$

Using matrix mechanics

$$\mathbf{L}_x = \frac{1}{2}(\mathbf{L}_+ + \mathbf{L}_-)$$

$$\langle 1\,1 | \mathbf{L}_x | 1\,0 \rangle = \langle 1\,1 | \frac{1}{2}(\mathbf{L}_+ + \mathbf{L}_-) | 1\,0 \rangle$$

$$= \frac{\hbar\sqrt{2}}{2} \left(\langle 1\,1 | 1\,1 \rangle + \langle 1\,1 | 1\,-1 \rangle \right)$$

$$= \frac{\hbar}{\sqrt{2}}$$

Proceeding in this manner, the entire matrix may be filled out:

$$\langle \ell m | \mathbf{L}_x | \ell m' \rangle =$$

	ℓ	\multicolumn{3}{c}{1}
ℓ	$m\backslash m'$	-1	0	$+1$
	-1	0	$\hbar/\sqrt{2}$	0
1	0	$\hbar/\sqrt{2}$	0	$\hbar/\sqrt{2}$
	$+1$	0	$\hbar/\sqrt{2}$	0

written more simply like

$$\mathbf{L}_x = \frac{\hbar}{\sqrt{2}} \begin{bmatrix} 0 & 1 & 0 \\ 1 & 0 & 1 \\ 0 & 1 & 0 \end{bmatrix}$$

Whenever an operator, like $\mathbf{L}_x$ above, is written as a matrix, it is assumed that you know the basis. In this case the basis is the usual one in which $\mathbf{L}^2$ and $\mathbf{L}_z$ have diagonal representations.

Clearly such a matrix has specified the operator L_x on a given subspace. The only thing preventing us from specifying this operator on its entire Hilbert space is that the matrix is infinite. For spin angular momentum, the matrices are finite.

Problem 1.5

The vector operator $\vec{L} = \hat{x}L_x + \hat{y}L_y + \hat{z}L_z$. Show that the matrix $\langle 1\,m\,|\vec{L}|\,1\,m'\rangle$ is given by

$m\backslash m'$	-1	0	$+1$
-1	$-\hbar\hat{z}$	$\frac{\hbar}{\sqrt{2}}(\hat{x}+i\hat{y})$	0
0	$\frac{\hbar}{\sqrt{2}}(\hat{x}-i\hat{y})$	0	$\frac{\hbar}{\sqrt{2}}(\hat{x}+i\hat{y})$
$+1$	0	$\frac{\hbar}{\sqrt{2}}(\hat{x}-i\hat{y})$	$\hbar\hat{z}$

All of the relations that have been developed for the orbital angular momentum $\vec{L}$ are valid for a general angular momentum operator $\vec{J}$. This notation explicitly assumes that there exists some angular momentum in quantum mechanics other than orbital angular momentum, as is indeed the case. In particular, the following relations define the angular momentum:

$$\text{If} \quad [J_i, J_j] = i\hbar\epsilon_{ijk}J_k$$

$$\text{And} \quad J^2|\,j\,m\rangle = \hbar^2 j\,(j+1)\,|\,j\,m\rangle$$

$$J_z|\,j\,m\rangle = \hbar m|\,j\,m\rangle$$

$$\text{Let} \quad J_+ = J_x + iJ_y$$

$$J_- = J_x - iJ_y$$

$$\text{Then} \quad [J_z, J_\pm] = \pm\hbar J_\pm$$

$$[J_+, J_-] = 2\hbar J_z$$

$$[J^2, J_\pm] = 0$$

$$J_+|\,j\,m\rangle = \hbar\,[(j-m)(j+m+1)]^{1/2}\,|\,j\,m+1\rangle$$

$$J_-|\,j\,m\rangle = \hbar\,[(j+m)(j-m+1)]^{1/2}\,|\,j\,m-1\rangle$$

There is one consequence of these equations yet to be examined. Consider

$$\mathbf{J}_+^p \,|\, j\, m\,\rangle \quad \text{with } p \text{ integer}$$

$$\text{if } m < j \quad \mathbf{J}_+ |\, j\, m\,\rangle \rightarrow \alpha |\, j\, m + 1\,\rangle$$
$$\mathbf{J}_+^2 |\, j\, m\,\rangle \rightarrow \alpha\beta |\, j\, m + 2\,\rangle$$
$$\mathbf{J}_+^p |\, j\, m\,\rangle \rightarrow \alpha\beta \ldots |\, j\, m + p\,\rangle$$
$$= \alpha\beta \ldots |\, j\, j\,\rangle$$

So $m + p = j; p = j - m$. Similarly,

$$\mathbf{J}_-^q |\, j\, m\,\rangle \rightarrow \alpha'\beta' \ldots |\, j\, m - q\,\rangle$$
$$= \alpha'\beta' \ldots |\, j\, -j\,\rangle$$

So $m - q = -j; q = j + m$. Therefore

$$p + q = 2j \quad \text{where } p + q \text{ are } \textit{integers.}$$

So $2j$ must be an integer and j can at most be half-integer. Of course it can also be integer.

The main points can be summarized as:

1. The only possible eigenvalues of $\mathbf{J}^2$ are of the form $j\,(j+1)$ where j is a nonnegative, integral, or half-integral number:

$$j = 0, \tfrac{1}{2}, 1, \tfrac{3}{2}, 2, \ldots, \infty$$

2. The only possible eigenvalues of $\mathbf{J}_z$ are the integral and half-integral numbers:

$$m = 0, \pm\tfrac{1}{2}, \pm 1, \pm\tfrac{3}{2}, \pm 2, \ldots, \pm\infty$$

3. If $j\,(j+1)$ and m are the respective eigenvalues of $\mathbf{J}^2$ and $\mathbf{J}_z$, then the only possible values of m are the $(2j + 1)$ quantities

$$-j, -j + 1, \ldots, j - 1, j$$

Based on experiments involving the Zeeman effect and the Stern–Gerlach experiment, it is postulated that an electron has an intrinsic angular momentum $\vec{\mathbf{S}}$ of magnitude $\hbar/2$ (spin $\tfrac{1}{2}$). That is, the existence of an observable operator $\vec{\mathbf{S}}$ is postulated along with an associated Hilbert space with two dimensions such that

$$\mathbf{S}^2 |\, s\, m_s\,\rangle = s(s+1)\hbar^2 |\, s\, m_s\,\rangle$$
$$\mathbf{S}_z |\, s\, m_s\,\rangle = m_s \hbar |\, s\, m_s\,\rangle$$

Now $s = \frac{1}{2}$, so

$$\mathbf{S}^2 | s\, m_s \rangle = \frac{3}{4} \hbar^2 | s\, m_s \rangle$$

and

$$| s\, m_s \rangle = | \frac{1}{2}\, \frac{1}{2} \rangle \quad \text{or} \quad | \frac{1}{2}\, -\frac{1}{2} \rangle \quad only$$

All of the previous formulae for a general angular momentum are applicable. In particular

$$\mathbf{S}_+ | \frac{1}{2}\, -\frac{1}{2} \rangle = \hbar | \frac{1}{2}\, \frac{1}{2} \rangle$$

$$\mathbf{S}_- | \frac{1}{2}\, +\frac{1}{2} \rangle = \hbar | \frac{1}{2}\, -\frac{1}{2} \rangle$$

$$\mathbf{S}_+ | \frac{1}{2}\, \frac{1}{2} \rangle = \mathbf{S}_- | \frac{1}{2}\, -\frac{1}{2} \rangle = 0$$

$$\mathbf{S}_- \mathbf{S}_+ | \frac{1}{2}\, -\frac{1}{2} \rangle = \hbar^2 | \frac{1}{2}\, -\frac{1}{2} \rangle$$

$$\mathbf{S}_+ \mathbf{S}_- | \frac{1}{2}\, \frac{1}{2} \rangle = \hbar^2 | \frac{1}{2}\, \frac{1}{2} \rangle$$

The smallness of the Hilbert space leads to some interesting relations. In what follows you need to consider the operator acting upon the basis kets (either one!):

$$\mathbf{S}_+^2 | s\, m_s \rangle = \mathbf{S}_-^2 | s\, m_s \rangle = 0$$

$$\mathbf{S}_x = \frac{1}{2}(\mathbf{S}_+ + \mathbf{S}_-) \tag{1.31}$$

$$\mathbf{S}_x^2 = \frac{1}{4}(\mathbf{S}_+^2 + \mathbf{S}_-^2 + \mathbf{S}_+\mathbf{S}_- + \mathbf{S}_-\mathbf{S}_+) \tag{1.32}$$

$$= \frac{1}{4}(\mathbf{S}_+\mathbf{S}_- + \mathbf{S}_-\mathbf{S}_+) \tag{1.33}$$

$$\mathbf{S}_x^2 | s\, m_s \rangle = \frac{1}{4}(\mathbf{S}_+\mathbf{S}_- + \mathbf{S}_-\mathbf{S}_+) | s\, m_s \rangle = \frac{\hbar^2}{4} | s\, m_s \rangle$$

for either state. Similarly (show these)

$$\mathbf{S}_y^2 | s\, m_s \rangle = \mathbf{S}_z^2 | s\, m_s \rangle = \frac{\hbar^2}{4} | s\, m_s \rangle$$

So $\mathbf{S}_x^2 = \mathbf{S}_y^2 = \mathbf{S}_z^2 = \frac{\hbar^2}{4}$.

Since $\mathbf{S}_+^2 = (\mathbf{S}_x + i\mathbf{S}_y)^2 = (\mathbf{S}_x^2 - \mathbf{S}_y^2) + i(\mathbf{S}_x\mathbf{S}_y + \mathbf{S}_y\mathbf{S}_x)$, it follows that $\mathbf{S}_x\mathbf{S}_y + \mathbf{S}_y\mathbf{S}_x = 0$. In fact, this relationship holds for any two of the three spin operators, that is, $\mathbf{S}_x$, $\mathbf{S}_y$ and $\mathbf{S}_z$ anticommute. This will be evident once we develop the matrix form for these operators:

$$\text{Anticommutator} \quad \equiv \{\mathbf{A}, \mathbf{B}\} \equiv \mathbf{AB} + \mathbf{BA}$$

It is easy to write down the matrices of $\vec{S}$ in the basis $|s\,m_s\rangle$. One proceeds exactly as was done for $\mathbf{L}_x$ in the basis $|1\,m_l\rangle$. Since our basis now consists of two vectors, our matrices are two dimensional. It is customary to define

$$\vec{S} \equiv \frac{\hbar}{2}\vec{\sigma}$$

and to work out the matrices for $\vec{\sigma}$, called the Pauli spin matrices. Using the basis $|s\,m_s\rangle$ with $s = \tfrac{1}{2}$, sometimes the bras and kets will be written using m_s alone. Let's do the one for σ_y:

$$\langle \tfrac{1}{2}\,|\sigma_y|\,-\tfrac{1}{2}\rangle = \langle \tfrac{1}{2}\,|\frac{-i}{\hbar}(\mathbf{S}_+ - \mathbf{S}_-)|\,-\tfrac{1}{2}\rangle = -i$$

The other three can be done in the same manner giving

$$\langle s\,m_s\,|\sigma_y|\,s\,m'_s\rangle =$$

$m\backslash m'$	$-\tfrac{1}{2}$	$\tfrac{1}{2}$
$-\tfrac{1}{2}$	0	i
$\tfrac{1}{2}$	$-i$	0

The shorthand form for these is

$$\sigma_x = \begin{bmatrix} 0 & 1 \\ 1 & 0 \end{bmatrix}$$

$$\sigma_y = \begin{bmatrix} 0 & -i \\ i & 0 \end{bmatrix}$$

$$\sigma_z = \begin{bmatrix} 1 & 0 \\ 0 & -1 \end{bmatrix}$$

and the basis is

$$|\tfrac{1}{2}\,\tfrac{1}{2}\rangle = \begin{pmatrix} 1 \\ 0 \end{pmatrix} \quad \text{sometimes called } \alpha$$

$$|\tfrac{1}{2}\,-\tfrac{1}{2}\rangle = \begin{pmatrix} 0 \\ 1 \end{pmatrix} \quad \text{sometimes called } \beta$$

You may verify

$$\sigma_\alpha^2 = \sigma_x^2 = \sigma_y^2 = \sigma_z^2 = 1 \qquad 1 = \begin{bmatrix} 1 & 0 \\ 0 & 1 \end{bmatrix}$$

$$\sigma_\alpha\sigma_\beta = \epsilon_{\alpha\beta\gamma}i\sigma_\gamma + \delta_{\alpha\beta}$$

$$\sigma_x\sigma_y\sigma_z = i \quad (\text{means } i1)$$

$$\mathrm{Tr}\sigma_x = \mathrm{Tr}\sigma_y = \mathrm{Tr}\sigma_z = 0$$

$$\det\sigma_x = \det\sigma_y = \det\sigma_z = -1$$

Note: The subscripts on σ have nothing to do with the matrix indices themselves. This is frequently a confusing point for students.

Now the wave functions for hydrogen should be modified to include spin. The notation changes from

$$| n \ell m_\ell \rangle \quad \text{an orbital}$$

$$\text{to} \quad | n \ell m_\ell m_s \rangle \equiv | n \ell m_\ell \rangle | {}^1\!/_2 \, m_s \rangle, \quad \text{called a spin orbital.}$$

The expression $| n \ell m_\ell \rangle | {}^1\!/_2 \, m_s \rangle$ is the tensor product of two Hilbert spaces having a dimension twice as large as $| n \ell m_\ell \rangle$ alone. The effect of any operator independent of the spin is trivial:

$$\langle n \ell m_\ell m_s | \mathbf{A} | n' \ell' m'_\ell m'_s \rangle = \langle n \ell m_\ell | \mathbf{A} | n' \ell' m'_\ell \rangle \langle m_s | m'_s \rangle$$

$$= \langle n \ell m_\ell | \mathbf{A} | n' \ell' m'_\ell \rangle \delta_{m_s m'_s}$$

So far, our Hamiltonian does not depend on the spin, so there would be no change to the energy levels of hydrogen with this modification (not true, however, for complex atoms). There is a small effect (fine structure) this change makes for hydrogen which will be developed subsequently.

1.6   Addition of Angular Momentum

Having introduced the concept of a generalized angular momentum and then having shown that spin satisfies all of the conditions of such an angular momentum, it remains to show that one can add two generalized angular momenta and obtain a generalized angular momentum. One can, for example, add orbital and spin angular momenta without being accused of adding apples and oranges. This is not a trivial point because electron spin is *not* an electron spinning on its axis; all such classical models fail. It is a purely quantum mechanical attribute of fundamental particles, not derivable from classical physics, which combines with an attribute that is derivable from classical physics. You may be forgiven for thinking that is amazing.

Consider

$$\vec{J} = \vec{J}_1 + \vec{J}_2 \tag{1.34}$$

where $\vec{J}_1$ and $\vec{J}_2$ refer to the angular momentum of two different systems or of noninteracting parts of one system which together form the system under study. Either way, the requirement is that

$$\left[\vec{J}_1, \vec{J}_2 \right] = 0. \tag{1.35}$$

By this is meant that each component of one angular momentum commutes with every component of the other.

$$\text{Then} \quad [\mathbf{J}_i, \mathbf{J}_j] = [\mathbf{J}_{1i} + \mathbf{J}_{2i}, \mathbf{J}_{1j} + \mathbf{J}_{2j}]$$

$$= [\mathbf{J}_{1i}, \mathbf{J}_{1j}] + [\mathbf{J}_{2i}, \mathbf{J}_{2j}]$$

$$= \epsilon_{ijk} i\hbar (\mathbf{J}_{1k} + \mathbf{J}_{2k}) \tag{1.36}$$

$$= \epsilon_{ijk} i\hbar \mathbf{J}_k.$$

$\vec{\mathbf{J}}$ can also be an angular momentum with $\mathbf{J}_\pm$ defined as before. Proofs of all the commutator relations are done as above. $\vec{\mathbf{J}}$ will satisfy all such relations.

What about eigenvalues of $\vec{\mathbf{J}}$ and the state vectors? To be an angular momentum, there must exist state vectors such that

$$\mathbf{J}^2 |\, j\, m\,\rangle = j\,(j+1)\,\hbar^2 |\, j\, m\,\rangle \tag{1.37}$$

$$\text{and} \quad \mathbf{J}_z |\, j\, m\,\rangle = m\hbar |\, j\, m\,\rangle. \tag{1.38}$$

What is j in terms of j_1 and j_2? What is $|\, j\, m\,\rangle$ in terms of $|\, j_1\, m_1\,\rangle$, $|\, j_2\, m_2\,\rangle$?

Consider the easily formed state

$$|\, j_1\, j_2\, m_1\, m_2\,\rangle \equiv |\, j_1\, m_1\,\rangle |\, j_2\, m_2\,\rangle. \tag{1.39}$$

This is a vector in a $(2j_1 + 1)(2j_2 + 1)$ dimensional Hilbert space. The vector $|\, j\, m\,\rangle$, more completely written as $|\, j_1\, j_2\, j\, m\,\rangle$, is a vector in the same space. There must exist a unitary transformation between these two sets defined as

$$|\, j_1\, j_2\, j\, m\,\rangle = \sum_{m_1 m_2} |\, j_1\, j_2\, m_1\, m_2\,\rangle \langle\, j_1\, j_2\, m_1\, m_2\, |\, j_1\, j_2\, j\, m\,\rangle. \tag{1.40}$$

The coefficients $\langle\, j_1\, j_2\, m_1\, m_2\, |\, j_1\, j_2\, j\, m\,\rangle$ are called Clebsch–Gordan coefficients (C–G's) or vector-coupling coefficients. Thought of as a matrix, they form a unitary matrix. However, the phases are chosen such that these coefficients are *real*.

Note that $\mathbf{J}_z$ commutes with $\mathbf{J}_{1z}$ and $\mathbf{J}_{2z}$. However, $\mathbf{J}^2$ does *not* commute with $\mathbf{J}_{1z}$ and $\mathbf{J}_{2z}$ separately (though it does commute with their sum). Furthermore $\mathbf{J}_1^2$ and $\mathbf{J}_2^2$ commute with $\mathbf{J}^2$, $\mathbf{J}_z$ as well as $\mathbf{J}_{1z}$ and $\mathbf{J}_{2z}$. So there are two different equivalent sets of commuting operators:

$$\text{either} \quad \mathbf{J}_1^2 \, \mathbf{J}_2^2 \, \mathbf{J}_{1z} \, \mathbf{J}_{2z}$$

$$\text{or} \quad \mathbf{J}_1^2 \, \mathbf{J}_2^2 \, \mathbf{J}^2 \, \mathbf{J}_z$$

This is the reason for adopting the notation above.

Next notice that

$$\mathbf{J}_z |\, j_1\, j_2\, j\, m\,\rangle = m\hbar |\, j_1\, j_2\, j\, m\,\rangle$$

$$\mathbf{J}_z |\, j_1\, j_2\, m_1\, m_2\,\rangle = (m_1 + m_2)\hbar |\, j_1\, j_2\, m_1\, m_2\,\rangle$$

So the set of eigenvalues for $\mathbf{J}_z$ can be readily found in either basis. Now the eigenvalues of an operator are independent of basis, so

$$m = m_1 + m_2. \tag{1.41}$$

Furthermore, using considerations of ladder operators it can be shown that

$$j_1 + j_2 \geq j \geq |j_1 - j_2| \tag{1.42}$$

The veracity of this relation can be demonstrated by reflecting on the sizes of the two Hilbert spaces and realizing that they must have the same dimension. This dimension is simply $(2j_1 + 1)(2j_2 + 1)$ in the j_1, j_2 basis. The dimension in the j basis is obtained by counting the degeneracy, $(2j + 1)$, for each allowed value of j. The following problem asks you to demonstrate this.

Problem 1.6

Show

$$\sum_{j=|j_1-j_2|}^{j_1+j_2} (2j + 1) = (2j_1 + 1)(2j_2 + 1)$$

A feel for C–G coefficients can best be obtained by example: consider $j_1 = 1$, $j_2 = \frac{1}{2}$, then $j = \frac{3}{2}$ or $\frac{1}{2}$. The vectors which span the space for each of the bases are

$\lvert j_1\, j_2\, j\, m \rangle$	$\lvert j_1\, j_2\, m_1\, m_2 \rangle$
$\lvert 1\,\frac{1}{2}\,\frac{3}{2}\,\frac{3}{2} \rangle$	$\lvert 1\,\frac{1}{2}\,1\,\frac{1}{2} \rangle$
$\lvert 1\,\frac{1}{2}\,\frac{3}{2}\,\frac{1}{2} \rangle$	$\lvert 1\,\frac{1}{2}\,0\,\frac{1}{2} \rangle$
$\lvert 1\,\frac{1}{2}\,\frac{1}{2}\,\frac{1}{2} \rangle$	$\lvert 1\,\frac{1}{2}\,1\,-\frac{1}{2} \rangle$
$\lvert 1\,\frac{1}{2}\,\frac{3}{2}\,-\frac{1}{2} \rangle$	$\lvert 1\,\frac{1}{2}\,-1\,\frac{1}{2} \rangle$
$\lvert 1\,\frac{1}{2}\,\frac{1}{2}\,-\frac{1}{2} \rangle$	$\lvert 1\,\frac{1}{2}\,0\,-\frac{1}{2} \rangle$
$\lvert 1\,\frac{1}{2}\,\frac{3}{2}\,-\frac{3}{2} \rangle$	$\lvert 1\,\frac{1}{2}\,-1\,-\frac{1}{2} \rangle$

The relation $m = m_1 + m_2$ is extremely powerful. It means the following kets *must* satisfy

$$\lvert 1\,\tfrac{1}{2}\,\tfrac{3}{2}\,\tfrac{3}{2} \rangle = \lvert 1\,\tfrac{1}{2}\,1\,\tfrac{1}{2} \rangle \tag{1.43}$$

$$\text{and} \quad \lvert 1\,\tfrac{1}{2}\,\tfrac{3}{2}\,-\tfrac{3}{2} \rangle = \lvert 1\,\tfrac{1}{2}\,-1\,-\tfrac{1}{2} \rangle. \tag{1.44}$$

The remaining four cluster into groups of two. The first group is

$$\lvert 1\,\tfrac{1}{2}\,\tfrac{3}{2}\,\tfrac{1}{2} \rangle = \alpha\lvert 1\,\tfrac{1}{2}\,0\,\tfrac{1}{2} \rangle + \beta\lvert 1\,\tfrac{1}{2}\,1\,-\tfrac{1}{2} \rangle \tag{1.45}$$

$$\lvert 1\,\tfrac{1}{2}\,\tfrac{1}{2}\,\tfrac{1}{2} \rangle = \gamma\lvert 1\,\tfrac{1}{2}\,0\,\tfrac{1}{2} \rangle + \delta\lvert 1\,\tfrac{1}{2}\,1\,-\tfrac{1}{2} \rangle \tag{1.46}$$

Because the kets on the right-hand side are orthogonal, one can multiply by their bras to solve for the four coefficients. The expressions one obtains can be compared to those in Eq. (1.40) which are Clebsch–Gordan coefficients:

$$\alpha = \langle 1\,\tfrac{1}{2}\,0\,\tfrac{1}{2}\,|\,1\,\tfrac{1}{2}\,\tfrac{3}{2}\,\tfrac{1}{2}\,\rangle$$

$$\beta = \langle 1\,\tfrac{1}{2}\,1\,-\tfrac{1}{2}\,|\,1\,\tfrac{1}{2}\,\tfrac{3}{2}\,\tfrac{1}{2}\,\rangle$$

$$\gamma = \langle 1\,\tfrac{1}{2}\,0\,\tfrac{1}{2}\,|\,1\,\tfrac{1}{2}\,\tfrac{1}{2}\,\tfrac{1}{2}\,\rangle \qquad \text{Clebsch–Gordan coefficients}$$

$$\delta = \langle 1\,\tfrac{1}{2}\,1\,-\tfrac{1}{2}\,|\,1\,\tfrac{1}{2}\,\tfrac{1}{2}\,\tfrac{1}{2}\,\rangle$$

The kets on the LHS of Eqs. (1.45) and (1.46) are orthonormal, so $\alpha^2 + \beta^2 = 1$, $\gamma^2 + \delta^2 = 1$, and $\alpha\gamma + \beta\delta = 0$.

One consistent way of finding C–G's is to start with the maximum vector and work down with ladder operators. Start with Eq. (1.43):

$$\mathbf{J}_- |\,1\,\tfrac{1}{2}\,\tfrac{3}{2}\,\tfrac{3}{2}\,\rangle = \sqrt{3}\hbar|\,1\,\tfrac{1}{2}\,\tfrac{3}{2}\,\tfrac{1}{2}\,\rangle$$

$$(\mathbf{J}_{1-} + \mathbf{J}_{2-})|\,1\,\tfrac{1}{2}\,1\,\tfrac{1}{2}\,\rangle = \sqrt{2}\hbar|\,1\,\tfrac{1}{2}\,0\,\tfrac{1}{2}\,\rangle + \hbar|\,1\,\tfrac{1}{2}\,1\,-\tfrac{1}{2}\,\rangle$$

Equating these two expressions immediately gives Eq. (1.45) from which the values may be obtained:

$$\alpha = \frac{\sqrt{2}}{\sqrt{3}} \qquad \beta = \frac{1}{\sqrt{3}}.$$

Orthonormality then gives

$$\gamma = \pm\frac{1}{\sqrt{3}} \qquad \delta = \mp\frac{\sqrt{2}}{\sqrt{3}}.$$

Choice of sign is a matter of convention. The chosen convention is that $\gamma = -1/\sqrt{3}$ and $\delta = +\sqrt{2}/\sqrt{3}$.

Problem 1.7

Find the C–G coefficients for the second group of two.

One important property of C–G's is that they are *real*. Therefore,

$$\langle j_1\,j_2\,m_1\,m_2\,|\,j_1\,j_2\,j\,m\,\rangle = \langle j_1\,j_2\,j\,m\,|\,j_1\,j_2\,m_1\,m_2\,\rangle \tag{1.47}$$

One consequence of deciding upon a certain phase for the C–G's is that the order of coupling the vectors is no longer arbitrary. Reversing the order introduces a phase factor:

$$\langle j_1\,j_2\,m_1\,m_2\,|\,j_1\,j_2\,j\,m\,\rangle = (-1)^{j_1+j_2-j}\langle j_2\,j_1\,m_2\,m_1\,|\,j_2\,j_1\,j\,m\,\rangle \tag{1.48}$$

Note: Some authors write $|\,j_1\,j_2\,m_1\,m_2\,\rangle$ as $|\,j_1\,m_1\,j_2\,m_2\,\rangle$, which is the same thing.

There are many symmetry relations exhibited by C–G coefficients. Many of them become simpler to write if one defines a new symbol. The $3-j$ symbol is defined as

$$\begin{pmatrix} j_1 & j_2 & j \\ m_1 & m_2 & m \end{pmatrix} = (-1)^{j_1-j_2-m}(2j+1)^{-1/2} \langle j_1 j_2 m_1 m_2 | j_1 j_2 j \ -m \rangle \quad (1.49)$$

These $3-j$ symbols have become a standard notation for angular momentum theory. Extensive use of them will not be made in this text, but we will invoke certain properties and make use of certain relations without trying to prove them:

1. An even permutation of columns leaves the value of a $3-j$ symbol unchanged, e.g.,

$$\begin{pmatrix} j_1 & j_2 & j_3 \\ m_1 & m_2 & m_3 \end{pmatrix} = \begin{pmatrix} j_2 & j_3 & j_1 \\ m_2 & m_3 & m_1 \end{pmatrix} \quad (1.50)$$

2. An odd permutation is equivalent to multiplication by $(-1)^{j_1+j_2+j_3}$, e.g.,

$$\begin{pmatrix} j_2 & j_1 & j_3 \\ m_2 & m_1 & m_3 \end{pmatrix} = (-1)^{j_1+j_2+j_3}\begin{pmatrix} j_1 & j_2 & j_3 \\ m_1 & m_2 & m_3 \end{pmatrix} \quad (1.51)$$

3. All of the signs of the m's can be changed:

$$\begin{pmatrix} j_1 & j_2 & j_3 \\ m_1 & m_2 & m_3 \end{pmatrix} = (-1)^{j_1+j_2+j_3}\begin{pmatrix} j_1 & j_2 & j_3 \\ -m_1 & -m_2 & -m_3 \end{pmatrix} \quad (1.52)$$

4. For the $3-j$ symbol $\begin{pmatrix} j_1 & j_2 & j_3 \\ m_1 & m_2 & m_3 \end{pmatrix}$ $m_1 + m_2 + m_3 = 0$ always.

From the symmetry relation (3) above, we deduce that

$$\begin{pmatrix} j_1 & j_2 & j_3 \\ 0 & 0 & 0 \end{pmatrix} = 0 \qquad \text{if } j_1 + j_2 + j_3 \text{ odd.}$$

Two orthogonality relations among $3-j$ symbols are particularly useful. These are

$$\sum_{j_3,m_3} (2j_3+1)\begin{pmatrix} j_1 & j_2 & j_3 \\ m_1 & m_2 & m_3 \end{pmatrix}\begin{pmatrix} j_1 & j_2 & j_3 \\ m_1' & m_2' & m_3 \end{pmatrix} = \delta_{m_1,m_1'}\delta_{m_2,m_2'} \quad (1.53)$$

$$\sum_{m_1, m_2} \begin{pmatrix} j_1 & j_2 & j_3 \\ m_1 & m_2 & m_3 \end{pmatrix} \begin{pmatrix} j_1 & j_2 & j_3' \\ m_1 & m_2 & m_3' \end{pmatrix} = \delta_{j_3, j_3'} \delta_{m_3, m_3'} (2j_3 + 1)^{-1} \quad (1.54)$$

There is one further relation that should be mentioned before leaving $3-j$ symbols. It is called the Wigner–Eckart theorem. This theorem states that if one forms the matrix element of an operator whose angular dependence is the same as the spherical harmonics between angular momentum states, then the m-dependence is given by a $3-j$ symbol independent of the original operator.

Operators that have the same angular dependence as the spherical harmonics are called *spherical tensors* or, more properly, *spherical tensor operators*. They are written as T_k^m and satisfy the same commutation relations with $\mathbf{J}_\pm$ and $\mathbf{J}_z$ as do the Y_ℓ^m. If one were to take some quantum mechanical operator, say the electric quadrupole moment, and express it as a spherical tensor, then the angular part of the problem becomes solved, and one is left with only a radial integral to perform. Formally the Wigner–Eckart theorem is expressed as

$$\boxed{\langle \gamma' j' m' | T_k^q | \gamma j m \rangle = (-1)^{j'-m'} \begin{pmatrix} j' & k & j \\ -m' & q & m \end{pmatrix} \langle \gamma' j' \| T_k \| \gamma j \rangle} \quad (1.55)$$

In the equation above, γ stands for all quantum numbers other than the ones relating to angular momentum. The symbol on the right is called a double-bar matrix element or reduced matrix element and is defined by this expression. It is evaluated by choosing the easiest matrix element on the left to perform. Once the double-bar matrix element is known, all other matrix elements involving different values for m, m', or q can be evaluated by just evaluating the $3-j$ symbol.

It is now possible for us to add the orbital and spin angular momentum of an electron in a hydrogen atom:

$$\vec{\mathbf{J}} = \vec{\mathbf{L}} + \vec{\mathbf{S}}$$

In the coupled basis there are the states $| n \ell s j m \rangle$ while in the uncoupled basis there are $| n \ell s m_l m_s \rangle \Rightarrow R_{n\ell}(r) Y_\ell^m(\theta, \phi) | \tfrac{1}{2} m_s \rangle$.

In QM only matrix elements are really important. Whether one can write down a wave function is not so important. Elementary treatments imply that to do the former, you must be able to do the latter. As this course develops, you will see that this is often not the case.

Consider the expectation value of an operator, $\mathbf{A}$, which does not depend on θ or ϕ and does *not* operate on the spin. Its expectation value in the coupled basis is desired:

$$\langle n\,\ell\,s\,j\,m\,|\mathbf{A}(r)|\,n\,\ell\,s\,j\,m\,\rangle$$

$$= \sum_{\substack{m'_\ell m'_s \\ m_\ell m_s}} \langle n\,\ell\,s\,j\,m\,|\,n\,\ell\,s\,m'_\ell\,m'_s\,\rangle \langle n\,\ell\,s\,m'_\ell\,m'_s\,|\mathbf{A}|\,n\,\ell\,s\,m_\ell\,m_s\,\rangle$$

$$\langle n\,\ell\,s\,m_\ell\,m_s\,|\,n\,\ell\,s\,j\,m\,\rangle$$

$$= \int R^*_{n\ell}(r)\mathbf{A}R_{n\ell}(r)r^2\,dr \sum_{\substack{m'_\ell m'_s \\ m_\ell m_s}} \delta_{m'_\ell m_\ell}\delta_{m'_s m_s}\langle j\,m\,|\,m'_\ell\,m'_s\,\rangle\langle m_\ell\,m_s\,|\,j\,m\,\rangle$$

$$= \int R^*_{n\ell}(r)\mathbf{A}R_{n\ell}(r)r^2\,dr \sum_{m_\ell m_s} \langle j\,m\,|\,m_\ell\,m_s\,\rangle\langle m_\ell\,m_s\,|\,j\,m\,\rangle$$

$$= \int R^*_{n\ell}(r)\mathbf{A}R_{n\ell}(r)r^2\,dr \qquad\qquad (1.56)$$

No Clebsch–Gordan coefficients appear in the final expression.

For one-electron systems, the notation in the coupled basis is built up in the following manner:

If one electron has $\ell = 0\ 1\ 2\ 3\ 4$
write $\quad s\ p\ d\ f\ g$

Then a complete description can be done by subscripting the letter with the j value. So,

$$s = {}^1\!/_2,\ \ell = 1,\ j = {}^3\!/_2 \text{ becomes } p_{3/2}$$
$$s = {}^1\!/_2,\ \ell = 1,\ j = {}^1\!/_2 \text{ becomes } p_{1/2}$$
$$s = {}^1\!/_2,\ \ell = 3,\ j = {}^5\!/_2 \text{ becomes } f_{5/2} \quad \text{etc.}$$

Initially it might seem confusing that the symbol s stands both for the spin quantum number in general and for the specific orbital quantum number of zero value. In practice, context will make clear which is meant. In spectroscopic notation, the m values are rarely given since except for applied external fields, all such states are degenerate.

1.7   Hamiltonian Consequences of Spin: Hydrogen Atom

It has already been mentioned that the addition of electron spin to hydrogen changes nothing since the Schrödinger equation is spin independent. When hydrogen is solved in the Dirac theory,[3] there appear additional terms, one of which looks like

$$\mathbf{H}_1 = \frac{-\hbar\mu_\circ}{mce} \frac{\vec{\mathbf{L}} \cdot \vec{\mathbf{S}}}{r} \frac{dV}{dr} \quad \text{in Gaussian units, } \vec{\mathbf{L}} \text{ and } \vec{\mathbf{S}} \text{ dimensionless}$$

Since $V = Ze^2/r$, this can be rewritten in atomic units as

$$\boxed{\mathbf{H}_1 = \frac{Z\alpha^2}{2} \frac{\vec{\mathbf{L}} \cdot \vec{\mathbf{S}}}{r^3}} \tag{1.57}$$

$\mathbf{H}_1$ is given in Hartrees. Since $\alpha = 1/137$, α^2 is a small quantity compared to the ground-state energy of $1/2$ Hartree. First-order perturbation theory may then be used to obtain the energy shifts, E_1, which can be used to correct the Schrödinger result for spin.

Recall whenever $\mathbf{H} = \mathbf{H}_0 + \mathbf{H}_1$ with $\mathbf{H}_1$ small, the exact result $\mathbf{H}\Psi = E\Psi$ can be approximated by

$$E = E_0 + E_1$$

where $\mathbf{H}_0\Psi_0 = E_0\Psi_0$ is the unperturbed solution and

$$E_1 = \int \Psi_0 \mathbf{H}_1 \Psi_0 \, d\tau. \tag{1.58}$$

In bra–ket notation this would be written

$$E_1 = \langle n \ell m_\ell m_s | \mathbf{H}_1 | n \ell m_\ell m_s \rangle \tag{1.59}$$

Now notice that $\vec{\mathbf{L}} \cdot \vec{\mathbf{S}} = 1/2[\mathbf{J}^2 - \mathbf{L}^2 - \mathbf{S}^2]$. This operator is diagonal in the basis $| n \ell s j m_j \rangle$. $\mathbf{H}_0$ is spin independent, so it too is diagonal in the $| n \ell s j m_j \rangle$ basis. Hence

$$E_1 = \langle n \ell s j m_j | \frac{Z\alpha^2}{4} \frac{[\mathbf{J}^2 - \mathbf{L}^2 - \mathbf{S}^2]}{r^3} | n \ell s j m_j \rangle$$

$$= \frac{Z\alpha^2}{4} [j(j+1) - \ell(\ell+1) - 3/4] \langle n \ell s j m_j | \frac{1}{r^3} | n \ell s j m_j \rangle. \tag{1.60}$$

But the expectation value on the right is given in Table 1.2:

---

[3]See *Quantum Mechanics of One- and Two-Electron Atoms*, Bethe and Salpeter, Plenum Press, 1977 (reprint of 1957).

$$\langle n\ell s j m_j | \frac{1}{r^3} | n\ell s j m_j \rangle = \int R^*_{n\ell}(r)\frac{1}{r^3}R_{n\ell}(r)r^2\,dr$$

$$= \frac{Z^3}{n^3\ell(\ell+\frac{1}{2})(\ell+1)}. \tag{1.61}$$

$$\text{Let}\quad \xi_{n\ell} \equiv \frac{\alpha^2 Z^4}{2n^3\ell(\ell+\frac{1}{2})(\ell+1)} \text{ Hartrees} \tag{1.62}$$

$$\text{then}\quad E_1 = \frac{1}{2}\,\xi_{n\ell}\,[j\,(j+1) - \ell\,(\ell+1) - \frac{3}{4}]. \tag{1.63}$$

Since there are additional terms in the Hamiltonian, this correction to the Schrödinger equation is not expected to be complete. We can, however, expect it to give the correct value for fine-structure splitting since this was the only spin-dependent term which appeared. The fine structure is the energy between the two states $j = \ell + \frac{1}{2}$ and $j = \ell - \frac{1}{2}$, for example, the energy between $2p_{3/2}$ and $2p_{1/2}$:

$$\Delta E_{fs} = E_1(j) - E_1(j-1) = j\xi_{n\ell} = (\ell+\frac{1}{2})\xi_{n\ell}$$

$$\Delta E_{fs} = \frac{\alpha^2 Z^4}{2n^3\ell\,(\ell+1)} \tag{1.64}$$

$$\Delta E_{fs} = \frac{R\alpha^2 Z^4}{n^3\ell\,(\ell+1)}$$

where R is the Rydberg constant. For hydrogen, $R\alpha^2 = 5.8405$ cm^{-1}. So the fine structure between the $2p_{3/2}$ and $2p_{1/2}$ (the largest such splitting in hydrogen) is

$$\Delta E(2p_{3/2} - 2p_{1/2}) = 0.365 \text{ cm}^{-1}$$

This is less than one part in 2×10^5 of the ground-state energy.

1.8   Useful Tables

The tables following will be used throughout the text and should be self-explanatory. A few comments here might be of some help to the student. The first table (Table 1.1), giving normalized hydrogenic radial wave functions, is not needed for the solution to any exercise or problem but is useful when using programs that use such functions so that one can ensure that normalization factors are those being assumed in this text. The second table (Table 1.2) that offers matrix elements of powers of r is the one that will be used when doing such matrix elements "by

Table 1.1 Normalized hydrogenic radial wave functions

n	ℓ	$rR_{n\ell}(r)$
1	0	$2Z^{\frac{3}{2}}\, r\, e^{(-Zr)}$
2	0	$\frac{1}{4}\sqrt{2}\, Z^{\frac{3}{2}}\, r\, e^{(-Zr/2)}(2 - Z\, r)$
2	1	$\frac{1}{12}\sqrt{6}\, Z^{\frac{5}{2}}\, r^2\, e^{(-Zr/2)}$
3	0	$\frac{2}{27}\sqrt{3}\, Z^{\frac{3}{2}}\, r\, e^{(-Zr/3)}\left(3 - 2\, Z\, r + \frac{2}{9}\, Z^2\, r^2\right)$
3	1	$\frac{1}{81}\sqrt{6}\, Z^{\frac{5}{2}}\, r^2\, e^{(-Zr/3)}\left(4 - \frac{2}{3}\, Z\, r\right)$
3	2	$\frac{2}{1215}\sqrt{30}\, Z^{\frac{7}{2}}\, r^3\, e^{(-Zr/3)}$
4	0	$\frac{1}{16}\, Z^{\frac{3}{2}}\, r\, e^{(-Zr/4)}\left(4 - 3\, Z\, r + \frac{1}{2}\, Z^2\, r^2 - \frac{1}{48}\, Z^3\, r^3\right)$
4	1	$\frac{1}{480}\sqrt{15}\, Z^{\frac{5}{2}}\, r^2\, e^{(-Zr/4)}\left(10 - \frac{5}{2}\, Z\, r + \frac{1}{8}\, Z^2\, r^2\right)$
4	2	$\frac{1}{1920}\sqrt{5}\, Z^{\frac{7}{2}}\, r^3\, e^{(-Zr/4)}\left(6 - \frac{1}{2}\, Z\, r\right)$
4	3	$\frac{1}{26880}\sqrt{35}\, Z^{\frac{9}{2}}\, r^4\, e^{(-Zr/4)}$
5	0	$\frac{2}{125}\sqrt{5}\, Z^{\frac{3}{2}}\, r\, e^{(-Zr/5)}\left(5 - 4\, Z\, r + \frac{4}{5}\, Z^2\, r^2 - \frac{4}{75}\, Z^3\, r^3 + \frac{2}{1875}\, Z^4\, r^4\right)$
5	1	$\frac{1}{1875}\sqrt{30}\, Z^{\frac{5}{2}}\, r^2\, e^{(-Zr/5)}\left(20 - 6\, Z\, r + \frac{12}{25}\, Z^2\, r^2 - \frac{4}{375}\, Z^3\, r^3\right)$
5	2	$\frac{2}{65625}\sqrt{70}\, Z^{\frac{7}{2}}\, r^3\, e^{(-Zr/5)}\left(21 - \frac{14}{5}\, Z\, r + \frac{2}{25}\, Z^2\, r^2\right)$
5	3	$\frac{1}{328125}\sqrt{70}\, Z^{\frac{9}{2}}\, r^4\, e^{(-Zr/5)}\left(8 - \frac{2}{5}\, Z\, r\right)$
5	4	$\frac{2}{4921875}\sqrt{70}\, Z^{\frac{11}{2}}\, r^5\, e^{(-Zr/5)}$
6	0	$\frac{1}{108}\sqrt{6}\, Z^{\frac{3}{2}}\, r\, e^{(-Zr/6)}\left(6 - 5\, Z\, r + \frac{10}{9}\, Z^2\, r^2 - \frac{5}{54}\, Z^3\, r^3 + \frac{1}{324}\, Z^4\, r^4 - \frac{1}{29160}\, Z^5\, r^5\right)$
6	1	$\frac{1}{11340}\sqrt{210}\, Z^{\frac{5}{2}}\, r^2\, e^{(-Zr/6)}\left(35 - \frac{35}{3}\, Z\, r + \frac{7}{6}\, Z^2\, r^2 - \frac{7}{162}\, Z^3\, r^3 + \frac{1}{1944}\, Z^4\, r^4\right)$
6	2	$\frac{1}{136080}\sqrt{105}\, Z^{\frac{7}{2}}\, r^3\, e^{(-Zr/6)}\left(56 - \frac{28}{3}\, Z\, r + \frac{4}{9}\, Z^2\, r^2 - \frac{1}{162}\, Z^3\, r^3\right)$
6	3	$\frac{1}{1224720}\sqrt{35}\, Z^{\frac{9}{2}}\, r^4\, e^{(-Zr/6)}\left(36 - 3\, Z\, r + \frac{1}{18}\, Z^2\, r^2\right)$
6	4	$\frac{1}{7348320}\sqrt{7}\, Z^{\frac{11}{2}}\, r^5\, e^{(-Zr/6)}\left(10 - \frac{1}{3}\, Z\, r\right)$
6	5	$\frac{1}{242494560}\sqrt{77}\, Z^{\frac{13}{2}}\, r^6\, e^{(-Zr/6)}$

hand." If one chooses to use a computer program to perform the calculation, one can compare to the table to make sure that results agree. The third table (Table 1.3), like the first, may not be used directly but is useful for comparison when using a computer program. The fourth table (Table 1.4), offering integrals of three spherical harmonic functions, is needed on occasion and will prove more useful for future reference than for solving problems in the text. Such an integral arises when calculating matrix elements of any operator expressed as a spherical tensor.

The last table has been taken from the *Atomic, Molecular and Optical Physics Handbook* of the American Physical Society (used with permission) and gives the ground-state configurations for each of the elements. In addition, it gives the spectroscopic term designation (described in the following chapter) for the ground level of each element and the ionization energy in electron volts (Table 1.5).

Table 1.2 Matrix elements for r^α

	$\int R_{n\ell}(r) r^\alpha R_{n'\ell'}(r) r^2 \, dr$
$n' = n$ and $\ell' = \ell$	
$\alpha = 4$	$\frac{1}{8} n^4 [63n^4 - 35n^2(2\ell^2 + 2\ell - 3) + 5\ell(\ell+1)(3\ell^2 + 3\ell - 10) + 12]$
$\alpha = 3$	$\frac{1}{8} n^2 [35n^2(n^2 - 1) - 30n^2(\ell+2)(\ell-1) + 3(\ell+2)(\ell+1)\ell(\ell-1)]$
$\alpha = 2$	$\frac{1}{2} n^2 [5n^2 + 1 - 3\ell(\ell+1)]$
$\alpha = 1$	$\frac{1}{2} [3n^2 - \ell(\ell+1)]$
$\alpha = -1$	$\frac{1}{n^2}$
$\alpha = -2$	$\frac{1}{n^3(\ell+\frac{1}{2})}$
$\alpha = -3$	$\frac{1}{n^3(\ell+1)(\ell+\frac{1}{2})\ell}$
$\alpha = -4$	$\frac{3n^2 - \ell(\ell+1)}{2n^5(\ell+\frac{3}{2})(\ell+1)(\ell+\frac{1}{2})\ell(\ell-\frac{1}{2})}$
$n' = n$ and $\ell' = \ell - 1$	
$\alpha = 1$	$\frac{3}{2} n [n^2 - \ell^2]^{\frac{1}{2}}$
$n > n' \; n' = 1$ and $\ell' = 0$	
$\alpha = \ell$	$2^{\ell+3} \frac{n^{\ell+2}(n-1)^{n-\ell-2}\ell}{(n+1)^{n+\ell+2}} \left[\frac{(n+\ell)!}{(n-\ell-1)!} \right]^{\frac{1}{2}}$
$\alpha = \ell - 1$	$2^{\ell+2} \frac{n^{\ell}(n-1)^{n-\ell-1}}{(n+1)^{n+\ell+1}} \left[\frac{(n+\ell)!}{(n-\ell-1)!} \right]^{\frac{1}{2}}$
$n > n' \; n' = 2$ and $\ell' = 1$	
$\alpha = \ell + 1$	$2^{3\ell+7} \frac{n^{\ell+4}(n-2)^{n-\ell-4}}{\sqrt{6}(n+2)^{n+\ell+4}} \left[\frac{(n+\ell)!}{(n-\ell-1)!} \right]^{\frac{1}{2}} (2\ell^3 - 3\ell^2 - 5\ell + 2)$
$\alpha = \ell$	$2^{3\ell+5} \frac{n^{\ell+2}(n-2)^{n-\ell-3}}{\sqrt{6}(n+2)^{n+\ell+3}} \left[\frac{(n+\ell)!}{(n-\ell-1)!} \right]^{\frac{1}{2}} [(2\ell^2 - 3\ell - 1)n^2 + 4(\ell+1)]$
$\alpha = \ell - 1$	$2^{3\ell+4} \frac{n^{\ell+2}(n-2)^{n-\ell-2}(\ell-1)}{\sqrt{6}(n+2)^{n+\ell+2}} \left[\frac{(n+\ell)!}{(n-\ell-1)!} \right]^{\frac{1}{2}}$
$\alpha = \ell - 2$	$2^{3\ell+2} \frac{(n-2)^{n-\ell-1}}{\sqrt{6}(n+2)^{n+\ell+1}} \left[\frac{(n+\ell)!}{(n-\ell-1)!} \right]^{\frac{1}{2}}$
$n > n' \; n' = 2$ and $\ell' = 0$	
$\alpha = \ell$	$2^{3\ell+5+\frac{1}{2}} \frac{n^{\ell+2}(n-2)^{n-\ell-3}\ell}{(n+2)^{n+\ell+3}} \left[\frac{(n+\ell)!}{(n-\ell-1)!} \right]^{\frac{1}{2}} [(-\ell+2)n^2 - 4]$

Table 1.3 Spherical harmonic functions, $Y_\ell^m(\theta, \phi)$

Y_0^0	$4\pi^{-\frac{1}{2}}$
$Y_1^{\pm 1}$	$\mp(3/8\pi)^{\frac{1}{2}} \sin\theta\, e^{\pm i\phi}$
Y_1^0	$(3/4\pi)^{\frac{1}{2}} \cos\theta$
$Y_2^{\pm 2}$	$(15/32\pi)^{\frac{1}{2}} \sin^2\theta\, e^{\pm 2i\phi}$
$Y_2^{\pm 1}$	$\mp(15/8\pi)^{\frac{1}{2}} \sin\theta\cos\theta\, e^{\pm i\phi}$
Y_2^0	$(5/4\pi)^{\frac{1}{2}} [(3/2)\cos^2\theta - 1/2]$
$Y_3^{\pm 3}$	$\mp(35/64\pi)^{\frac{1}{2}} \sin^3\theta\, e^{\pm 3i\phi}$
$Y_3^{\pm 2}$	$(105/32\pi)^{\frac{1}{2}} \sin^2\theta\cos\theta\, e^{\pm 2i\phi}$
$Y_3^{\pm 1}$	$\mp(21/64\pi)^{\frac{1}{2}} \sin\theta[5\cos^2\theta - 1]\, e^{\pm i\phi}$
Y_3^0	$(7/4\pi)^{\frac{1}{2}} [(5/2)\cos^3\theta - (3/2)\cos\theta]$
$Y_4^{\pm 4}$	$(3/8)(35/8\pi)^{\frac{1}{2}} \sin^4\theta\, e^{\pm 4i\phi}$
$Y_4^{\pm 3}$	$\mp(3/4)(35/4\pi)^{\frac{1}{2}} \sin^3\theta\cos\theta\, e^{\pm 3i\phi}$
$Y_4^{\pm 2}$	$(3/4)(5/8\pi)^{\frac{1}{2}} \sin^2\theta[7\cos^2\theta - 1]\, e^{\pm 2i\phi}$
$Y_4^{\pm 1}$	$\mp(3/4)(5/4\pi)^{\frac{1}{2}} \sin\theta[7\cos^3\theta - 3\cos\theta]\, e^{\pm i\phi}$
Y_4^0	$(9/4\pi)^{\frac{1}{2}} [(35/8)\cos^4\theta - (15/4)\cos^2\theta + (3/8)]$

Table 1.4 Integral of three spherical harmonic functions

$\int Y_\ell^{m*}(\theta, \phi) Y_{\ell'}^{m'}(\theta, \phi) Y_\beta^\gamma(\theta, \phi) d(\sin\theta) d\phi$

$\beta = 1 \; \gamma = 0$

$\ell' = \ell + 1$ $\qquad\qquad (-1)^{\ell+1} \left[\dfrac{3(\ell+m+1)(\ell-m+1)}{4\pi(2\ell+3)(2\ell+1)} \right]^{1/2}$

$m' = m$

$\beta = 1 \; \gamma = \pm 1$

$\ell' = \ell + 1$ $\qquad\qquad (-1)^{\ell} \left[\dfrac{3(\ell-m+1)(\ell-m+2)}{8\pi(2\ell+3)(2\ell+1)} \right]^{1/2}$

$m' = m \mp 1$

$\beta = 2 \; \gamma = 0$

$\ell' = \ell + 2$ $\qquad\qquad (-1)^{\ell} \dfrac{3}{4(2\ell+3)} \left[\dfrac{5(\ell+m+2)(\ell+m+1)(\ell-m+1)(\ell-m+2)}{\pi(2\ell+5)(2\ell+1)} \right]^{1/2}$

$m' = m$

$\ell' = \ell$ $\qquad\qquad (-1)^{\ell} \dfrac{\sqrt{5}}{2\sqrt{\pi}} \dfrac{3m^2 - \ell(\ell+1)}{(2\ell+3)(2\ell-1)}$

$m' = m$

$\beta = 2 \; \gamma = \pm 1$

$\ell' = \ell + 2$ $\qquad\qquad (-1)^{\ell+1} \dfrac{1}{2(2\ell+3)} \left[\dfrac{15(\ell\pm m+1)(\ell\mp m+3)(\ell\mp m+2)(\ell\mp m+1)}{2\pi(2\ell+5)(2\ell+1)} \right]^{1/2}$

$m' = m \mp 1$

$\ell' = \ell$ $\qquad\qquad (-1)^{\ell+1} \dfrac{(1-2m)[15(\ell\mp m+1)(\ell\pm m)]^{1/2}}{2(2\pi)^{1/2}(2\ell+3)(2\ell-1)}$

$m' = m \mp 1$

$\beta = 2 \; \gamma = \pm 2$

$\ell' = \ell + 2$ $\qquad\qquad (-1)^{\ell} \dfrac{1}{4(2\ell+3)} \left[\dfrac{15(\ell\mp m+1)(\ell\mp m+2)(\ell\mp m+3)(\ell\mp m+4)}{2\pi(2\ell+5)(2\ell+1)} \right]^{1/2}$

$m' = m \mp 2$

$\ell' = \ell$ $\qquad\qquad (-1)^{\ell} \dfrac{[15(\ell\pm m-1)(\ell\pm m)(\ell\mp m+1)(\ell\mp m+2)]^{1/2}}{2(2\pi)^{1/2}(2\ell+3)(2\ell-1)}$

$m' = m \mp 2$

Table 1.5 Ground levels and ionization energies for the neutral atoms

Z	Element	Ground configuration[a]	Ground level	Ionization energy (eV)	Z	Element	Ground configuration[a]	Ground level	Ionization energy (eV)
1	H	$1s$	$^2S_{1/2}$	13.5984	53	I	$[\mathrm{Kr}]\,4d^{10}\,5s^2\,5p^5$	$^2P^\circ_{3/2}$	10.4513
2	He	$1s^2$	$^1S_0$	24.5874	54	Xe	$[\mathrm{Kr}]\,4d^{10}\,5s^2\,5p^6$	$^1S_0$	12.1298
3	Li	$1s^2\,2s$	$^2S_{1/2}$	5.3917	55	Cs	$[\mathrm{Xe}]\,6s$	$^2S_{1/2}$	3.8939
4	Be	$1s^2\,2s^2$	$^1S_0$	9.3227	56	Ba	$[\mathrm{Xe}]\,6s^2$	$^1S_0$	5.2117
5	B	$1s^2\,2s^2\,2p$	$^2P^\circ_{1/2}$	8.2980	57	La	$[\mathrm{Xe}]\,5d\,6s^2$	$^2D_{3/2}$	5.5770
6	C	$1s^2\,2s^2\,2p^2$	$^3P_0$	11.2603	58	Ce	$[\mathrm{Xe}]\,4f\,5d\,6s^2$	$^1G^\circ_4$	5.5387
7	N	$1s^2\,2s^2\,2p^3$	$^4S^\circ_{3/2}$	14.5341	59	Pr	$[\mathrm{Xe}]\,4f^3\,6s^2$	$^4I^\circ_{9/2}$	5.464
8	O	$1s^2\,2s^2\,2p^4$	$^3P_2$	13.6181	60	Nd	$[\mathrm{Xe}]\,4f^4\,6s^2$	$^5I_4$	5.5250
9	F	$1s^2\,2s^2\,2p^5$	$^2P^\circ_{3/2}$	17.4228	61	Pm	$[\mathrm{Xe}]\,4f^5\,6s^2$	$^6H^\circ_{5/2}$	5.58
10	Ne	$1s^2\,2s^2\,2p^6$	$^1S_0$	21.5646	62	Sm	$[\mathrm{Xe}]\,4f^6\,6s^2$	$^7F_0$	5.6436
11	Na	$[\mathrm{Ne}]\,3s$	$^2S_{1/2}$	5.1391	63	Eu	$[\mathrm{Xe}]\,4f^7\,6s^2$	$^8S^\circ_{7/2}$	5.6704
12	Mg	$[\mathrm{Ne}]\,3s^2$	$^1S_0$	7.6462	64	Gd	$[\mathrm{Xe}]\,4f^7\,5d\,6s^2$	$^9D^\circ_2$	6.1501
13	Al	$[\mathrm{Ne}]\,3s^2\,3p$	$^2P^\circ_{1/2}$	5.9858	65	Tb	$[\mathrm{Xe}]\,4f^9\,6s^2$	$^6H^\circ_{15/2}$	5.8638
14	Si	$[\mathrm{Ne}]\,3s^2\,3p^2$	$^3P_0$	8.1517	66	Dy	$[\mathrm{Xe}]\,4f^{10}\,6s^2$	$^5I_8$	5.9389
15	P	$[\mathrm{Ne}]\,3s^2\,3p^3$	$^4S^\circ_{3/2}$	10.4867	67	Ho	$[\mathrm{Xe}]\,4f^{11}\,6s^2$	$^4I^\circ_{15/2}$	6.0215
16	S	$[\mathrm{Ne}]\,3s^2\,3p^4$	$^3P_2$	10.3600	68	Er	$[\mathrm{Xe}]\,4f^{12}\,6s^2$	$^3H_6$	6.1077
17	Cl	$[\mathrm{Ne}]\,3s^2\,3p^5$	$^2P^\circ_{3/2}$	12.9676	69	Tm	$[\mathrm{Xe}]\,4f^{13}\,6s^2$	$^2F^\circ_{7/2}$	6.1843
18	Ar	$[\mathrm{Ne}]\,3s^2\,3p^6$	$^1S_0$	15.7596	70	Yb	$[\mathrm{Xe}]\,4f^{14}\,6s^2$	$^1S_0$	6.2542
19	K	$[\mathrm{Ar}]\,4s$	$^2S_{1/2}$	4.3407	71	Lu	$[\mathrm{Xe}]\,4f^{14}\,5d\,6s^2$	$^2D_{3/2}$	5.4259
20	Ca	$[\mathrm{Ar}]\,4s^2$	$^1S_0$	6.1132	72	Hf	$[\mathrm{Xe}]\,4f^{14}\,5d^2\,6s^2$	$^3F_2$	6.8251
21	Sc	$[\mathrm{Ar}]\,3d\,4s^2$	$^2D_{3/2}$	6.5615	73	Ta	$[\mathrm{Xe}]\,4f^{14}\,5d^3\,6s^2$	$^4F_{3/2}$	7.5496
22	Ti	$[\mathrm{Ar}]\,3d^2\,4s^2$	$^3F_2$	6.8281	74	W	$[\mathrm{Xe}]\,4f^{14}\,5d^4\,6s^2$	$^5D_0$	7.8640
23	V	$[\mathrm{Ar}]\,3d^3\,4s^2$	$^4F_{3/2}$	6.7463	75	Re	$[\mathrm{Xe}]\,4f^{14}\,5d^5\,6s^2$	$^6S_{5/2}$	7.8335
24	Cr	$[\mathrm{Ar}]\,3d^5\,4s$	$^7S_3$	6.7665	76	Os	$[\mathrm{Xe}]\,4f^{14}\,5d^6\,6s^2$	$^5D_4$	8.4382
25	Mn	$[\mathrm{Ar}]\,3d^5\,4s^2$	$^6S_{5/2}$	7.4340	77	Ir	$[\mathrm{Xe}]\,4f^{14}\,5d^7\,6s^2$	$^4F_{9/2}$	8.9670
26	Fe	$[\mathrm{Ar}]\,3d^6\,4s^2$	$^5D_4$	7.9024	78	Pt	$[\mathrm{Xe}]\,4f^{14}\,5d^9\,6s$	$^3D_3$	8.9587
27	Co	$[\mathrm{Ar}]\,3d^7\,4s^2$	$^4F_{9/2}$	7.8810	79	Au	$[\mathrm{Xe}]\,4f^{14}\,5d^{10}\,6s$	$^2S_{1/2}$	9.2255
28	Ni	$[\mathrm{Ar}]\,3d^8\,4s^2$	$^3F_4$	7.6398	80	Hg	$[\mathrm{Xe}]\,4f^{14}\,5d^{10}\,6s^2$	$^1S_0$	10.4375
29	Cu	$[\mathrm{Ar}]\,3d^{10}\,4s$	$^2S_{1/2}$	7.7264	81	Tl	$[\mathrm{Xe}]\,4f^{14}\,5d^{10}\,6s^2\,6p$	$^2P^\circ_{1/2}$	6.1082
30	Zn	$[\mathrm{Ar}]\,3d^{10}\,4s^2$	$^1S_0$	9.3942	82	Pb	$[\mathrm{Xe}]\,4f^{14}\,5d^{10}\,6s^2\,6p^2$	$^3P_0$	7.4167
31	Ga	$[\mathrm{Ar}]\,3d^{10}\,4s^2\,4p$	$^2P^\circ_{1/2}$	5.9993	83	Bi	$[\mathrm{Xe}]\,4f^{14}\,5d^{10}\,6s^2\,6p^3$	$^4S^\circ_{3/2}$	7.2856
32	Ge	$[\mathrm{Ar}]\,3d^{10}\,4s^2\,4p^2$	$^3P_0$	7.8994	84	Po	$[\mathrm{Xe}]\,4f^{14}\,5d^{10}\,6s^2\,6p^4$	$^3P_2$	8.4167
33	As	$[\mathrm{Ar}]\,3d^{10}\,4s^2\,4p^3$	$^4S^\circ_{3/2}$	9.7886	85	At	$[\mathrm{Xe}]\,4f^{14}\,5d^{10}\,6s^2\,6p^5$	$^2P^\circ_{3/2}$	
34	Se	$[\mathrm{Ar}]\,3d^{10}\,4s^2\,4p^4$	$^3P_2$	9.7524	86	Rn	$[\mathrm{Xe}]\,4f^{14}\,5d^{10}\,6s^2\,6p^6$	$^1S_0$	10.7485
35	Br	$[\mathrm{Ar}]\,3d^{10}\,4s^2\,4p^5$	$^2P^\circ_{3/2}$	11.8138	87	Fr	$[\mathrm{Rn}]\,7s$	$^2S_{1/2}$	4.0727
36	Kr	$[\mathrm{Ar}]\,3d^{10}\,4s^2\,4p^6$	$^1S_0$	13.9996	88	Ra	$[\mathrm{Rn}]\,7s^2$	$^1S_0$	5.2784
37	Rb	$[\mathrm{Kr}]\,5s$	$^2S_{1/2}$	4.1771	89	Ac	$[\mathrm{Rn}]\,6d\,7s^2$	$^2D_{3/2}$	5.17
38	Sr	$[\mathrm{Kr}]\,5s^2$	$^1S_0$	5.6949	90	Th	$[\mathrm{Rn}]\,6d^2\,7s^2$	$^3F_2$	6.3067
39	Y	$[\mathrm{Kr}]\,4d\,5s^2$	$^2D_{3/2}$	6.2171	91	Pa	$[\mathrm{Rn}]\,5f^2\,(^3H_4)\,6d\,7s^2$	$(4,\tfrac{3}{2})^\circ_{11/2}$	5.89
40	Zr	$[\mathrm{Kr}]\,4d^2\,5s^2$	$^3F_2$	6.6339	92	U	$[\mathrm{Rn}]\,5f^3\,(^4I^\circ_{9/2})\,6d\,7s^2$	$(^9/_2,^3/_2)^\circ_6$	6.1941
41	Nb	$[\mathrm{Kr}]\,4d^4\,5s$	$^6D_{1/2}$	6.7589	93	Np	$[\mathrm{Rn}]\,5f^4\,(^5I_4)\,6d\,7s^2$	$(4,\tfrac{3}{2})_{11/2}$	6.2657
42	Mo	$[\mathrm{Kr}]\,4d^5\,5s$	$^7S_3$	7.0924	94	Pu	$[\mathrm{Rn}]\,5f^6\,7s^2$	$^7F_0$	6.0262
43	Tc	$[\mathrm{Kr}]\,4d^5\,5s^2$	$^6S_{5/2}$	7.28	95	Am	$[\mathrm{Rn}]\,5f^7\,7s^2$	$^8S^\circ_{7/2}$	5.9738
44	Ru	$[\mathrm{Kr}]\,4d^7\,5s$	$^5F_5$	7.3605	96	Cm	$[\mathrm{Rn}]\,5f^7\,6d\,7s^2$	$^9D^\circ_2$	6.02
45	Rh	$[\mathrm{Kr}]\,4d^8\,5s$	$^4F_{9/2}$	7.4589	97	Bk	$[\mathrm{Rn}]\,5f^9\,7s^2$	$^6H^\circ_{15/2}$	6.23
46	Pd	$[\mathrm{Kr}]\,4d^{10}$	$^1S_0$	8.3369	98	Cf	$[\mathrm{Rn}]\,5f^{10}\,7s^2$	$^5I_8$	6.30
47	Ag	$[\mathrm{Kr}]\,4d^{10}\,5s$	$^2S_{1/2}$	7.5763	99	Es	$[\mathrm{Rn}]\,5f^{11}\,7s^2$	$^4I^\circ_{15/2}$	6.42
48	Cd	$[\mathrm{Kr}]\,4d^{10}\,5s^2$	$^1S_0$	8.9938	100	Fm	$[\mathrm{Rn}]\,5f^{12}\,7s^2$	$^3H_6$	6.50
49	In	$[\mathrm{Kr}]\,4d^{10}\,5s^2\,5p$	$^2P^\circ_{1/2}$	5.7864	101	Md	$[\mathrm{Rn}]\,5f^{13}\,7s^2$	$^2F^\circ_{7/2}$	6.58
50	Sn	$[\mathrm{Kr}]\,4d^{10}\,5s^2\,5p^2$	$^3P_0$	7.3439	102	No	$[\mathrm{Rn}]\,5f^{14}\,7s^2$	$^1S_0$	6.65
51	Sb	$[\mathrm{Kr}]\,4d^{10}\,5s^2\,5p^3$	$^4S^\circ_{3/2}$	8.6084	103	Lr	$[\mathrm{Rn}]\,5f^{14}\,7s^2\,7p$?	$^2P^\circ_{1/2}$?	4.9 ?
52	Te	$[\mathrm{Kr}]\,4d^{10}\,5s^2\,5p^4$	$^3P_2$	9.0096	104	Rf	$[\mathrm{Rn}]\,5f^{14}\,6d^2\,7s^2$?	$^3F_2$?	6.0 ?

[a] An element symbol in brackets represents the electrons in the ground configuration of that element.

Chapter 2
Complex Atoms

The multi-electron, central force problem is one that does not have an exact solution. Approximations must be applied, and some aspects of these approximations are common to all multi-particle problems, particularly those of nuclear physics. Atoms can rightly be thought of as the building blocks of our material world. Understanding how quantum mechanics describes atoms, which justifies so much that you have been taught in chemistry and modern physics courses, is the goal for the rest of this text. In this chapter, the hardest part of that broad effort will be attempted which is trying to understand the energy-level structure of an isolated, multi-electron atom. That will be what is meant by the "solution" to the problem at hand. The problem, as presented, will rapidly grow in complexity and seem to be impossible to handle. But then the complexity will shrink as so many terms of interest are shown to be equal to others or zero.

The problem being considered is to find the total energy of all of the electrons of a multi-electron atom under the influence of the Coulomb attraction of the electrons to the nucleus and the mutual repulsion of each of the other electrons. Solving the Schrödinger equation is not the approach to take. Rather one starts by evaluating the energy assuming that you know the wave functions. The variational procedure tells you that if you then modify the assumed wave functions in any way at all that lowers the energy, both the wave functions and energy are closer to the "correct" ones. Interestingly, how one performs this procedure is not really very important. The optimum method is the Hartree–Fock procedure which will be briefly described in the last section of this chapter. What is important is understanding how one evaluates the energy, what approximations are used in making that evaluation, and which quantum numbers can be used to describe the states available to the atom. After all, when doing atomic physics research, the atom is often in an excited state, so these are every bit as important to understand as the ground state of the atom.

R.L. Brooks, *The Fundamentals of Atomic and Molecular Physics*, Undergraduate Lecture Notes in Physics, DOI 10.1007/978-1-4614-6678-9_2,
© Springer Science+Business Media New York 2013

2.1   Shell Model of the Atom

Consider forming a system of several electrons electrostatically bound to an infinitely heavy nucleus of charge Ze. The electrons also electrostatically repel each other, so a Hamiltonian for the system can be written that, in atomic units, looks like

$$\mathbf{H} = \sum_{i=1}^{N} \left(-\frac{\nabla_i^2}{2} - \frac{Z}{|\vec{r}_i|} \right) + \sum_{i>j=1}^{N} \frac{1}{|\vec{r}_i - \vec{r}_j|}$$

$$\text{or} \quad \mathbf{H} = \sum_{i=1}^{N} \left(-\frac{\nabla_i^2}{2} - \frac{Z}{r_i} \right) + \sum_{i>j=1}^{N} \frac{1}{r_{ij}} \tag{2.1}$$

where $\vec{r}_i$ is the position of electron i with respect to the nucleus. r_i and r_{ij} are defined by the corresponding expression in the equation above. While the Hamiltonian is not complete, all additional terms can quite successfully be treated as perturbations.

The Schrödinger equation for this Hamiltonian is hopelessly complicated to solve using even the largest computer presently imaginable. The starting point for calculations is then taken to be the assumption of a *central potential*, $V(r_i)$, for each of the electrons. Various approximations differ by what is chosen for $V(r_i)$. Very often V is chosen *iteratively*: assume a V, calculate the wave functions, find a new V, and calculate new wave functions until the procedure converges, called the *self-consistent field* approximation. This treatment doesn't take that approach. Our potential really will be the one given above. However, the overall form of the wave functions, taken to be obtained as products of one-electron wave functions, is justified by looking at the form the wave functions would have if the potential were central.

Using a central potential $\mathbf{H}$ may be written as

$$\mathbf{H} = \sum_{i=1}^{N} -\frac{\nabla_i^2}{2} + V(r_i). \tag{2.2}$$

The Schrödinger equation is

$$\mathbf{H}u = Eu$$

$$\text{where} \quad u = \prod_{i=1}^{N} u_i(a_i) = u_1(a_1)u_2(a_2)\ldots u_N(A_N) \tag{2.3}$$

$$\text{and} \quad E = \sum_{i=1}^{N} E_i(a_i) \tag{2.4}$$

Here $u_i(a_i)$ is the wave function of the ith electron having a set of quantum numbers a_i.

Problem 2.1

Show that the above product wave function is a solution of the Schrödinger equation.

The separation of the wave function into a product of one-electron wave functions is a direct consequence of the central potential assumption and is the cornerstone of the atomic theory of complex atoms. In fact our potential is not a central potential, so how can such a separation be justified? The answer is that the largest part of the potential, attraction to the nucleus, is rigorously central and the electron–electron repulsion, which is not central, appears to average out enough to let this crucial approximation work amazingly well. Keep in mind that no change to the *potential* is being made here. This is an assumption relating to the overall form of the solution. Hence it will be assumed that the form of the wave function can be taken as a product of one- electron wave functions.

It follows then that each u_i is a solution of the one-electron central force problem and must be of the form

$$u_i(a_i) = \mathrm{R}_{n_i \ell_i}(r_i) \mathrm{Y}_{\ell_i}^{m_i}(\theta_i, \phi_i) \chi_{s_i}(i) \tag{2.5}$$

which is a one-electron spin orbital. The form of the $\mathrm{R}_{n_i \ell_i}(r_i)$ is unspecified and depends on the particular choice of central potential. So a_i must represent the set of four quantum numbers:

$$a_i \equiv n_i\, \ell_i\, m_{\ell_i}\, m_{s_i} \tag{2.6}$$

which could also be written as

$$u_i(a_i) = |\, n_i\, \ell_i\, m_{\ell_i}\, m_{s_i}\, \rangle. \tag{2.7}$$

What if the positions of the ith and jth electron were exchanged? The expression would be

$$u_j(a_i) = \mathrm{R}_{n_i \ell_i}(r_j) \mathrm{Y}_{\ell_i}^{m_i}(\theta_j, \phi_j) \chi_{s_i}(j)$$

So something is wrong with our solution, (2.3). It has erroneously assigned a given set of quantum numbers to a given electron *when the electrons cannot be distinguished*.

For brevity consider a 2-electron system. (2.3) might be more properly written as

$$u^s = u_1(a_1)u_2(a_2) + u_2(a_1)u_1(a_2) \tag{2.8}$$

$$u^a = u_1(a_1)u_2(a_2) - u_2(a_1)u_1(a_2). \tag{2.9}$$

u^s is symmetric under exchange of electrons, while u^a is antisymmetric (changes sign) under such an interchange. Both solutions satisfy the indistinguishability of electrons.

To explain the building-up of the periodic table of elements, Pauli postulated that no two electrons could have the same set of quantum numbers. Using this criterion the solution u^a is chosen over u^s since the former vanishes whenever the two electrons have the same quantum numbers. The correct generalization of the Pauli principle is that a system composed of fermions (spin $\frac{1}{2}$ odd integer) has a wave function which is antisymmetric against exchange of any two particles.

One could now define a symmetrization operator and an antisymmetrization operator which would form the necessary linear combinations to account for indistinguishability of particles. Since in atomic physics only the latter needs to be considered, there is a particularly elegant way to perform this. It is known as the Slater determinant. Let

$$u(a) \equiv \frac{1}{\sqrt{N!}} \begin{vmatrix} u_1(a_1) & u_1(a_2) & u_1(a_3) & \dots & u_1(a_N) \\ u_2(a_1) & u_2(a_2) & u_2(a_3) & \dots & u_2(a_N) \\ \vdots & & & & \\ u_N(a_1) & u_N(a_2) & u_N(a_3) & \dots & u_N(a_N) \end{vmatrix} \quad (2.10)$$

Note that this reduces to (2.9) for two particles. Also note that our first attempt as a solution, (2.3), is represented by the diagonal entries.

A more compact notation for $|u(a)\rangle$ is

$$|u(a)\rangle = \mathcal{A}|n_1\,\ell_1\,m_{\ell_1}\,m_{s_1}\rangle|n_2\,\ell_2\,m_{\ell_2}\,m_{s_2}\rangle \dots |n_N\,\ell_N\,m_{\ell_N}\,m_{s_N}\rangle$$

where $\mathcal{A}$ means antisymmetrical product or Slater determinant. Let us write

$$\begin{aligned} n = 3 & \quad \ell = 2 & \quad m_\ell = 1 & \quad m_s = \frac{1}{2} & \quad \text{as} & \quad |3\,d\,1^+\rangle \\ n = 3 & \quad \ell = 2 & \quad m_\ell = -2 & \quad m_s = -\frac{1}{2} & \quad \text{as} & \quad |3\,d\text{-}2^-\rangle \end{aligned}$$

which shall define our notation.

If there were four electrons in a $1s^2 2p 3d$ configuration, what might the wave function be? Four electrons require the specification of 16 quantum numbers whereas $1s^2 2p 3d$ specifies only eight of them. There must be a whole host of states permitted by the classification $1s^2 2p 3d$. *One* of them could look like

$$|u(a)\rangle = \mathcal{A}|1\,s\,0^+\rangle|1\,s\,0^-\rangle|2\,p\,1^+\rangle|3\,d\text{-}2^+\rangle$$

$$= \frac{1}{\sqrt{4!}} \begin{vmatrix} |1\,s\,0^+\rangle_1 & |1\,s\,0^-\rangle_1 & |2\,p\,1^+\rangle_1 & |3\,d\text{-}2^+\rangle_1 \\ |1\,s\,0^+\rangle_2 & |1\,s\,0^-\rangle_2 & |2\,p\,1^+\rangle_2 & |3\,d\text{-}2^+\rangle_2 \\ |1\,s\,0^+\rangle_3 & |1\,s\,0^-\rangle_3 & |2\,p\,1^+\rangle_3 & |3\,d\text{-}2^+\rangle_3 \\ |1\,s\,0^+\rangle_4 & |1\,s\,0^-\rangle_4 & |2\,p\,1^+\rangle_4 & |3\,d\text{-}2^+\rangle_4 \end{vmatrix}$$

The subscript to the right of each ket indicates the electron number. Each electron takes a turn getting each set of quantum numbers. There are four electrons and four specified sets of QN's, but one cannot say which electron has which set. Therefore

$|u(a)\rangle$ specified above is one state allowed to the configuration $1s^22p3d$. How many different states are there?

If a determinant were formed in which two of the states were $|1s0^+\rangle|1s0^+\rangle$ or $|1s0^-\rangle|1s0^-\rangle$, then two columns of the Slater determinant would be identical, and the state would be zero. Also, if the state $|1s0^-\rangle|1s0^+\rangle$ were formed, it would be the determinant previously written with the first two columns interchanged, yielding precisely the same state. So the specification $|1s0^+\rangle|1s0^-\rangle$ for $1s^2$ is unique.

Six different kets could be formed using a $2p$ configuration. They would be

$$|2p1^+\rangle, |2p1^-\rangle, |2p0^+\rangle, |2p0^-\rangle, |2p\text{-}1^+\rangle, |2p\text{-}1^-\rangle$$

Similarly, ten different kets could be formed using $3d$. All in all 60 different Slater determinants could be formed from the configuration $1s^22p3d$. Do these 60 different states yield 60 different energies, the same energy 60 times, or what?

The answer is that if the central potential were Z/r_i, all 60 states would be degenerate. The occurrence of $1/r_{ij}$ in the Hamiltonian partially lifts the degeneracy. (There are six different energies for these 60 states using the Hamiltonian of (2.1). This point will be returned to later.)

The shell structure of an atom should now be understandable. Only *one state* can be formed from the configuration $1s^2$. Similarly the configuration $1s^22s^2$ has only one state. It is formed by the determinant

$$|u(a)\rangle = \mathcal{A}|1s0^+\rangle|1s0^-\rangle|2s0^+\rangle|2s0^-\rangle$$

In this way all shells (those having the same value of n) when filled have only *one state*. $2n^2$ electrons fill a shell. The following are closed shells:

$$1s^2$$

$$2s^22p^6$$

$$3s^23p^63d^{10}$$

etc.

It is equally clear, however, that whenever $2(2\ell+1)$ electrons are assigned to an $n\ell$ subshell, there is only one state:

$$2s^2$$

$$3s^2 \quad \text{or} \quad 3p^6$$

$$4d^{10}$$

etc.

Multiple states for a single configuration arise only with partially filled subshells. All filled shells (given n value) or subshells (given n and ℓ values) have a total angular momentum of zero and are completely defined by the configuration specifications.

Problem 2.2

Consider all different filled subshells in the periodic table up to $_{54}$Xe. Which ones yield chemically nonreactive substances? Which are chemically reactive? Why?

2.2 Angular Momentum for Complex Atoms

In Chap. 1, the general rules for coupling two angular momenta were presented. One always couples angular momenta in pairs, usually starting from the innermost to the outermost electron in a partially filled subshell. (Filled subshells have zero total angular momentum.) The most common coupling scheme for atoms is called Russell–Saunders or LS coupling. In this scheme all of the orbital angular momenta are coupled together, all of the spin angular momenta are coupled together, and the two sums are then coupled together to form the total angular momentum, $\vec{\mathbf{J}}$:

$$\vec{\mathbf{L}} = \sum_{i=1}^{N} \vec{\ell}_i$$

$$\vec{\mathbf{S}} = \sum_{i=1}^{N} \vec{s}_i$$

$$\vec{\mathbf{J}} = \vec{\mathbf{L}} + \vec{\mathbf{S}}$$

For two electrons, everything carries over from the coupling of two angular momenta. So,

$$|\ell_1 - \ell_2| \le L \le |\ell_1 + \ell_2|$$

$$|s_1 - s_2| \le S \le |s_1 + s_2|$$

$$|L - S| \le J \le |L + S|$$

Even for more than two electrons, one can write

$$M_L = \sum_{i=1}^{N} m_{\ell_i}$$

$$M_S = \sum_{i=1}^{N} m_{s_i}$$

$$M_J = M_L + M_S$$

For two electrons in the configuration $2p3p$, the possible values for the total orbital angular momentum are $L = 0$, 1, or 2 whose symbols are S, P, or D. Since the spin for each is $s_i = 1/2$, the total spin can be $S = 1$ or 0. These possibilities are expressed as

$$^1S, \,^3S, \,^1P, \,^3P, \,^1D, \,^3D \quad \text{called } terms.$$

The rule is

$$^{2S+1}L_J$$

where $(2S + 1)$ is called the *multiplicity*. The *levels* can then be

$$^1S_0, \,^3S_1, \,^1P_1, \,^3P_{0,1,2}, \,^1D_2, \,^3D_{1,2,3}$$

where $^3P_{0,1,2}$ means $^3P_0$ and $^3P_1$ and $^3P_2$ pronounced "triplet pee two."

Proceeding as before, remember that the configuration $2p3p$ has 36 states, each with its own Slater determinant. The coupling done above effectively changes the basis from these 36 states to another set of 36 states, each of which is a linear combination of one or more of the original ones. Each total J value has $(2J + 1)$ different M_J values. The number of different M_J values for all of the above levels must be 36.

Problem 2.3

Verify that there are 36 states by counting the number of different M_J values for each of the levels given above.

For more than two electrons the angular momenta are coupled in pairs starting with the innermost configuration. The first pair then gives rise to all terms allowed by the angular momentum coupling rules. The next electron's angular momentum is then added to these terms, called parent terms, to obtain a new set of terms. Coupling angular momenta for four or more electrons from open subshells becomes more complicated and is treated in books on angular momentum theory.

The configuration $2p3p3d$ can yield a lot of terms. For example,

$$2p3p(^1S)3d \,^2D_{3/2,5/2}$$

$$2p3p(\underbrace{^3P}_{parent \text{ term}})3d \,^{2,4}P, \,^{2,4}D, \,^{2,4}F \Rightarrow \,^4F_{3/2,5/2,7/2,9/2}$$

This is only a selection of all possible terms that one could form. As a review of the terminology, consider where the second subscript is the M_J value.

an expression like ^2F is called a *term*

$^2F_{5/2}$ is called a *level*

$^2F_{5/2,5/2}$ is called a *state*

Do not confuse the notation $^2F_{5/2\,3/2}$ with $^2F_{7/2,\,5/2}$. The first has $J = {}^5/_2$ and $M_J = {}^3/_2$ and is a state designation. The second is a shorthand for the two levels $^2F_{7/2}$ and $^2F_{5/2}$.

The configuration $2p3p3d$ has 360 different states, each of which has a Slater determinant in the basis $|\,n_i\,\ell_i\,m_{\ell_i}\,m_{s_i}\,\rangle$. Coupling of the angular momentum represents a change of basis. In general, obtaining the transformation is complicated. Very often, only two or three electrons are outside closed subshells, which provides some simpler cases for consideration.

Problem 2.4

Write down all possible terms and levels from the configuration $2p3p3d$. Verify that in the coupled basis, there are 360 states. *Hint:* One must form all possible parent terms by coupling the inner two electrons before coupling the outer electron to each of these in turn.

The final coupling of $\vec{J} = \vec{L} + \vec{S}$ occurs only if the Hamiltonian has a term like $\vec{L} \cdot \vec{S}$ in it. This coupling is the weakest one and shall be ignored for the time being. Without such coupling, L, S, M_L, and M_S are good quantum numbers, and the state designations could be expressed as

$$^{(2S+1)}L_{M_L M_S}.$$

It will be made clear whenever such notation is used that the subscripts are M_L and M_S and not J and M_J. Again, this simply represents a different basis, actually a slightly easier one than JM_J. For a given term, the number of allowed M_L and M_S values is the same as the number of JM_J values.

Problem 2.5

Verify the claim made in the previous sentence.

Finally, note that the parity of a given state is given by $(-1)^{\Sigma \ell_i}$, *not by* $(-i)^L$. Why so? Because any state is a linear combination of Slater determinants made up of the product of q spin orbitals where q is the number of electrons. Each spin orbital has a parity determined by Y_ℓ^m which is $(-1)^\ell$. The parity of a product of q of them is then

$$(-1)^{\sum_{i=1}^q \ell_i}$$

The parity of any state is determined by the electron configuration and not by the coupling scheme.

The notation introduced in this section is particularly important for everything that follows. In particular, the distinction between the total spin, S, and the multiplicity, $2S + 1$, is often missed by many students.

2.3 Equivalent Electrons

You have seen that a lot of terms result from the coupling of three angular momenta. But our example chose three orbitals that were all different. What if there were three electrons in the *same* orbital? The Pauli exclusion principle insists that every electron have a unique set of quantum numbers and the use of Slater determinants ensures that. Whenever more than one electron resides in the same orbital, the electrons are called equivalent. This is the situation now being considered.

As previously mentioned, whenever two electrons are in an s^2 configuration, only a single state is formed. This follows since there is only one Slater determinant which is nonzero, given by

$$\frac{1}{\sqrt{2}} \begin{vmatrix} |s0^+\rangle_1 & |s0^-\rangle_1 \\ |s0^+\rangle_2 & |s0^-\rangle_2 \end{vmatrix}$$

where the unspecified quantum number n does not matter. The same argument holds for any filled subshell. But what happens to a partially filled subshell like $2p^2$? (Any p will do; the n value doesn't matter.) There is a clever tabular method which enables one to work out not only the number of allowed states (fewer than 36, the number for two nonequivalent p electrons) but also the allowed term values under LS coupling:

M_L	M_S	m_{ℓ_1}	m_{s_1}	m_{ℓ_2}	m_{s_2}	
2	1	1	$+1/2$	1	$+1/2$	Excluded by PP
2	0	1	$+1/2$	1	$-1/2$	OK
2	0	1	$-1/2$	1	$+1/2$	Same as above
2	-1	1	$-1/2$	1	$-1/2$	Excluded by PP
$\vdots$	$\vdots$	etc.		$\vdots$	$\vdots$	

One constructs a table in which each line represents one possible state available to the electrons in a given subshell. Each electron gets an entry designated by its m_ℓ and m_s values. The sum of these values for all of the electrons in the table is written to the left under the column headings of M_L, and M_S. Now reflect for a moment that each electron has four possible quantum numbers, n, ℓ, m_ℓ and m_s. By specifying only the m_ℓ and m_s numbers, there is a tacit assumption that n and ℓ are unimportant. The reason is that this table is only useful when all of the n and ℓ values are the same for each electron. If either of these values were different for two different electrons, those would not be equivalent, and the coupling could be performed as done previously.

A table for two nonequivalent p electrons would contain 36 entries since each electron can have six different m_ℓ, m_s combinations. Such a table is not particularly useful. However, when two of the p electrons are equivalent, many of these 36 entries will be excluded by the Pauli principle. A table for only the allowed combinations of m_{ℓ_1}, m_{s_1}, m_{ℓ_2}, and m_{s_2} would look like the following:

#	M_L	M_S	m_{ℓ_1}	m_{s_1}	m_{ℓ_2}	m_{s_2}
1	2	0	1	$+\frac{1}{2}$	1	$-\frac{1}{2}$
2	1	1	1	$+\frac{1}{2}$	0	$+\frac{1}{2}$
3	1	0	1	$+\frac{1}{2}$	0	$-\frac{1}{2}$
4	1	0	1	$-\frac{1}{2}$	0	$+\frac{1}{2}$
5	1	-1	1	$-\frac{1}{2}$	0	$-\frac{1}{2}$
6	0	1	1	$+\frac{1}{2}$	-1	$+\frac{1}{2}$
7	0	0	1	$+\frac{1}{2}$	-1	$-\frac{1}{2}$
8	0	0	1	$-\frac{1}{2}$	-1	$+\frac{1}{2}$
9	0	0	0	$+\frac{1}{2}$	0	$-\frac{1}{2}$
10	0	-1	1	$-\frac{1}{2}$	-1	$-\frac{1}{2}$
11	-1	1	-1	$+\frac{1}{2}$	0	$+\frac{1}{2}$
12	-1	0	-1	$+\frac{1}{2}$	0	$-\frac{1}{2}$
13	-1	0	-1	$-\frac{1}{2}$	0	$+\frac{1}{2}$
14	-1	-1	-1	$-\frac{1}{2}$	0	$-\frac{1}{2}$
15	-2	0	-1	$+\frac{1}{2}$	-1	$-\frac{1}{2}$

Any other combination you might think of is either zero upon forming a Slater determinant or is equivalent to one already written since the electrons are indistinguishable.

The terms of two inequivalent p electrons are

Term:	^1S	^3S	^1P	^3P	^1D	^3D
Degeneracy:	1	3	3	9	5	15

How many terms are possible for the case of $2p^2$? If any state belonging to a given term appears in the table, then *all* the states of that term must appear. It is best to start consideration from the largest value of M_L. Here that value is 2, and the largest value of M_S associated with that is 0, so there must be a ^1D. By count, 5 of the 15 entries in the table above are associated with this term. The next largest value of M_L is 1, and there are four entries in the table. The ^1D is responsible for one of these, one of the $M_L = 1$, $M_S = 0$. But there are three others, with $M_S = 1, 0$, and -1. A ^3P gives rise to those entries and 6 others besides, so 14 of the 15 entries are accounted for. There are three entries with $M_L = 0$ and $M_S = 0$, and the two terms found so far account for 2 of them. The third must belong to a ^1S, which accounts for only this entry and in total all 15 entries have been accounted for.

Note that if there were more than two electrons, the final terms that can occur depend on parentage, that is, the terms that are formed coupling in pairs from the inside out. For example, for a configuration of $2p3p4p$, there are 21 possible terms. This raises a point worth pursuing. Angular momentum is coupled in pairs, and it is important to keep track of the intermediate quantum numbers. The coupling of $2p$ and $3p$ leads to six terms specified previously. The ℓ and s values for the $4p$ electron need to be added to these six in turn. This procedure can be expressed as

$$2p3p(^1S)4p\ ^2P$$

$$2p3p(^3S)4p\ ^2P\ ^4P$$

$$2p3p(^1P)4p\ ^2S\ ^2P\ ^2D$$

$$2p3p(^3P)4p\ ^2S\ ^4S\ ^2P\ ^4P\ ^2D\ ^4D$$

$$2p3p(^1D)4p\ ^2P\ ^2D\ ^2F$$

$$2p3p(^3D)4p\ ^2P\ ^4P\ ^2D\ ^4D\ ^2F\ ^4F$$

In total, there are 21 terms not 8. It is important to realize that ^2P occurs six times. Each of those six has different energies and a recognizably different parentage. The same is true for many of the others. When being asked to give the terms that result from coupling three or more angular momenta, one must include the intermediate terms as part of the designation.

Problem 2.6

Show that the only allowed terms for a p^3 configuration are ^4S, ^2P, and ^2D. How many states are there? If the configuration were $2p3p4p$, how many states would there be?

2.4   Matrix Elements of the Hamiltonian

For elementary QM problems the "solution" will yield energy values (eigenvalues of the Hamiltonian) and wave functions simultaneously. One might think of such a solution as a "single-pass" solution. Problems that require iterative techniques for their solution need to evaluate the expectation value of the Hamiltonian as one step in a multi-pass algorithm. The variation technique, backbone for most approximate methods, is such an example. Besides this reason, there is the likelihood that the wave functions have been found for some approximate Hamiltonian (say, central potential) and that once these wave functions are known, you then desire to find the energy levels of the actual Hamiltonian. Either way one needs to know how to evaluate the matrix elements of the Hamiltonian between determinantal functions:

$$\text{Let} \quad \mathbf{H} = \sum_i \mathbf{f}_i + \sum_{\text{pairs } i,j} \mathbf{g}_{ij}$$

$$\text{where} \quad \mathbf{f}_i = -\frac{\nabla_i^2}{2} - \frac{Z}{r_i}$$

$$\text{and} \quad \mathbf{g}_{ij} = \frac{1}{r_{ij}}$$

Here, because $\mathbf{f}_i$ is a function of the coordinates of a single electron, it is called a one-electron operator, and $\mathbf{g}_{ij}$, having coordinates of two different electrons, is a two-electron operator.

The first step is to write the matrix element $\langle u\,|\mathbf{A}|\,u'\rangle$ of any operator $\mathbf{A}$ between determinantal functions:

$$\frac{1}{N!}\int \begin{vmatrix} u_1^*(1) & \cdots & u_1^*(N) \\ & \vdots & \\ u_N^*(1) & \cdots & u_N^*(N) \end{vmatrix} \mathbf{A} \begin{vmatrix} u_1'(1) & \cdots & u_1'(N) \\ & \vdots & \\ u_N'(1) & \cdots & u_N'(N) \end{vmatrix} d\tau_1 \ldots d\tau_N \qquad (2.11)$$

Here $u_1(1)$ is short for

$$u_1(a_1) = u_1(n_1, \ell_1, m_{\ell_1}, m_{s_1}) = R_{n_1\ell_1}(r_1)Y_{\ell_1}^{m_{\ell_1}}(\theta_1, \phi_1)\chi_{m_{s_1}}(1)$$

$$u_2(a_1) = R_{n_1\ell_1}(r_2)Y_{\ell_1}^{m_{\ell_1}}(\theta_2, \phi_2)\chi_{m_{s_1}}(2)$$

So the subscript on u is the electron index which goes with the coordinates (r, θ, ϕ) and thus with $d\tau_1$, $d\tau_2$, etc. ($d\tau_1 \equiv r_1^2 \sin\theta_1\, dr_1\, d\theta_1\, d\phi_1$.) This notation for $u_i(j)$ is backward from Slater's[1] but consistent with what was introduced previously. Now consider the term arising from the principal diagonal on the left:

$$\frac{1}{N!}\int u_1^*(1)u_2^*(2)\ldots u_N^*(N)\mathbf{A} \begin{vmatrix} u_1'(1) & \cdots & u_1'(N) \\ & \vdots & \\ u_N'(1) & \cdots & u_N'(N) \end{vmatrix} d\tau_1 \ldots d\tau_N \qquad (2.12)$$

Another term from the determinant on the left might be

$$\frac{1}{N!}\int -u_1^*(2)u_2^*(1)u_3^*(3)\ldots u_N^*(N)\mathbf{A} \begin{vmatrix} u_1'(1) & \cdots & u_1'(N) \\ & \vdots & \\ u_N'(1) & \cdots & u_N'(N) \end{vmatrix} d\tau_1 \ldots d\tau_N$$

Since the integration is performed over all electron coordinates, these are dummy indices. The above is the same as

$$\frac{1}{N!}\int -u_2^*(2)u_1^*(1)u_3^*(3)\ldots u_N^*(N)\mathbf{A} \begin{vmatrix} u_2'(1) & \cdots & u_2'(N) \\ u_1'(1) & \cdots & u_1'(N) \\ & \vdots & \\ u_N'(1) & \cdots & u_N'(N) \end{vmatrix} d\tau_1 \ldots d\tau_N$$

---

[1] John C. Slater, *Quantum Theory of Matter*, $2^n d$ ed., McGraw-Hill, New York, 1968.

Upon interchange of the first two rows of the determinant on the right, Eq. (2.12) is regained. In this way it can be shown that all $N!$ terms from the determinant on the left are the same as (2.12) which leads to

$$\langle\, u\,|\mathbf{A}|\, u'\,\rangle = \int u_1^*(1)\dots u_N^*(N)\mathbf{A} \begin{vmatrix} u_1'(1) & \dots & u_1'(N) \\ & \vdots & \\ u_N'(1) & \dots & u_N'(N) \end{vmatrix} \mathrm{d}\tau_1 \dots \mathrm{d}\tau_N$$

Further simplification depends on the form of the operator. It could be a constant, a one-electron operator, or a two-electron operator. Consider what happens if the operator is just a complex number. All u's are orthonormal, so unless one permutation matches the term on the left, the result is zero. A match can occur only if $u = u'$ for all N. Thus, unity results from the principal diagonal and zero from all other terms so that

$$\langle\, u\,|c|\, u'\,\rangle = c\langle\, u\,|\, u'\,\rangle = c \quad \text{for} \quad u = u'$$

$$= 0 \quad \text{for} \quad u \neq u'$$

2.4.1 One-Electron Operators

Let $\mathbf{A}$ be any one-electron operator $\mathbf{f}_i$. There are three cases to consider. Either u' is identical to u (one case) or it is different. If different, it can differ in any number of spin orbitals. However, should two or more spin orbitals be different, the resultant matrix element must be zero. After presenting the generality, a specific example should make the following cases clearer:

Case 1: $u = u'$ (diagonal matrix elements). Only the principal diagonal of the determinant on the right contributes yielding

$$\langle\, u\,|\sum_i \mathbf{f}_i|\, u\,\rangle = \int u_1^*(1)\mathbf{f}_1 u_1(1)\,\mathrm{d}\tau_1 + \int u_2^*(2)\mathbf{f}_2 u_2(2)\,\mathrm{d}\tau_2 + \dots$$

$$+ \int u_N^*(N)\mathbf{f}_N u_N(N)\,\mathrm{d}\tau_N$$

But $\mathbf{f}_i$ has the same functional form for each electron and the electron indices are dummy, so this expression may be written

$$\langle\, u\,|\sum_i \mathbf{f}_i|\, u\,\rangle = \sum_i \int u_1^*(i)\mathbf{f}_1 u_1(i)\,\mathrm{d}\tau_1$$

$$\equiv \sum_i \langle\, i\,|\mathbf{f}|\, i\,\rangle$$

Don't be confused by the index "1." When we started this all of the electrons were numbered. Now that number is redundant (and will shortly be dropped), so the "1" could equally well have been any of the electron numbers:

Case 2: $u = u'$ except for *one* spin orbital. Only a single term survives:

$$\langle i | \mathbf{f} | i' \rangle = \int u_1^*(i) \mathbf{f}_1 u_1'(i') \, d\tau_1$$

Case 3: $u = u'$ except for *more* than one spin orbital.

Zero.

Before proceeding to two-electron operators, working through a specific example should clarify what has been presented. Consider the configuration $1s2s2p$, three nonequivalent electrons. Choose the wave function (arbitrarily) to be

$$| u \rangle = \mathcal{A} | 1s0^+ \rangle | 2s0^+ \rangle | 2p1^- \rangle$$

This is one of 24 possible determinants from the configuration $1s2s2p$. Written out explicitly, the wave function looks like

$$u = \frac{1}{\sqrt{6}} \begin{vmatrix} |1s0^+\rangle_1 & |2s0^+\rangle_1 & |2p1^-\rangle_1 \\ |1s0^+\rangle_2 & |2s0^+\rangle_2 & |2p1^-\rangle_2 \\ |1s0^+\rangle_3 & |2s0^+\rangle_3 & |2p1^-\rangle_3 \end{vmatrix}$$

Let $\mathbf{f} = 1/r_i$, for example. Then

$$\langle u | \sum_i \mathbf{f}_i | u \rangle = \sum_i \langle (1s0^+)_1 (2s0^+)_2 (2p1^-)_3 | \frac{1}{r_i} |$$

$$| \begin{vmatrix} |1s0^+\rangle_1 & |2s0^+\rangle_1 & |2p1^-\rangle_1 \\ |1s0^+\rangle_2 & |2s0^+\rangle_2 & |2p1^-\rangle_2 \\ |1s0^+\rangle_3 & |2s0^+\rangle_3 & |2p1^-\rangle_3 \end{vmatrix} \rangle$$

The first term in the sum over i has six contributions from the determinant on the right. The one from the principal diagonal (the only nonzero one) is

$$\langle 1s0^+ | \frac{1}{r_1} | 1s0^+ \rangle_1 \langle 2s0^+ | 2s0^+ \rangle_2 \langle 2p1^- | 2p1^- \rangle_3$$

One of the five contributions not from the principal diagonal is

$$\langle 1s0^+ | \frac{1}{r_1} | 2s0^+ \rangle_1 \underbrace{\langle 2s0^+ | 1s0^+ \rangle_2}_{=0 \text{ by orthogonality}} \langle 2p1^- | 2p1^- \rangle_3$$

In this way all contributions from terms not of the principal diagonal are zero because at least one overlap integral (an integral not containing the operator) is zero.

Performing the sum over i would then yield

$$\langle u \mid \sum_i \mathbf{f}_i \mid u \rangle = \langle 1s0^+ \mid \frac{1}{r_1} \mid 1s0^+ \rangle_1 + \langle 2s0^+ \mid \frac{1}{r_2} \mid 2s0^+ \rangle_2$$

$$+ \langle 2p1^- \mid \frac{1}{r_3} \mid 2p1^- \rangle_3$$

Next notice that $1/r_1$, $1/r_2$, and $1/r_3$ have the same functional form. Only the electron index is different. But this difference is now no longer relevant because the integrals are isolated in individual terms of the sum. Hence they may be written as

$$\langle u \mid \sum_i \mathbf{f}_i \mid u \rangle = \langle 1s0^+ \mid \frac{1}{r_1} \mid 1s0^+ \rangle_1 + \langle 2s0^+ \mid \frac{1}{r_1} \mid 2s0^+ \rangle_1$$

$$+ \langle 2p1^- \mid \frac{1}{r_1} \mid 2p1^- \rangle_1$$

Now clearly the subscript 1 is irrelevant, and there is no possibility of confusion or ambiguity if the expression is written as

$$\langle u \mid \sum_i \mathbf{f}_i \mid u \rangle = \langle 1s0^+ \mid \frac{1}{r} \mid 1s0^+ \rangle + \langle 2s0^+ \mid \frac{1}{r} \mid 2s0^+ \rangle + \langle 2p1^- \mid \frac{1}{r} \mid 2p1^- \rangle.$$

Problem 2.7

Do you know what these integrals are for hydrogenic wave functions? Note that the sum over i has become a sum over quantum numbers.

Off-diagonal matrix elements are nonzero only if u' differs from u by not more than one spin orbital. This is quite a simplification for something that started off looking so formidable. For the one-electron operator $1/r$, it happens that there are no nonzero off-diagonal elements at all between different states of the same configuration. This happens because these different states differ only in their angular or spin dependence and the operator $1/r$ is insensitive to such dependence. But for a general one-electron operator, $\mathbf{f}(r, \theta, \phi)$, there can be nonzero elements.

Take another of our 24 possible determinants and write

$$\mid u' \rangle = \mathcal{A} \mid 1s0^+ \rangle \mid 2s0^+ \rangle \mid 2p0^- \rangle$$

Then $\langle u \mid \sum_i \mathbf{f}_i \mid u' \rangle$ has only the single nonzero contribution

$$\langle 1\,s\,0^+\,|\,1\,s\,0^+\rangle_1 \langle 2\,s\,0^+\,|\,2\,s\,0^+\rangle_2 \langle 2\,p\,1^-\,|\mathbf{f}_3(r,\theta,\phi)|\,2\,p\,0^-\rangle$$

or

$$\langle u\,|\sum_i \mathbf{f}_i\,|\,u'\rangle = \langle 2\,p\,1^-\,|\mathbf{f}(r,\theta,\phi)|\,2\,p\,0^-\rangle$$

Of course, even for a general one-electron operator, if it is not spin dependent, then the spins must be aligned, or the matrix element is zero, e.g.,

$$\langle u\,|\sum_i \mathbf{f}_i\,|\,u'\rangle = \langle 2\,p\,1^-\,|\mathbf{f}(r,\theta,\phi)|\,2\,p\,1^+\rangle = 0$$

because of spin orthogonality.

2.4.2 Two-Electron Operators

Let

$$\mathbf{A} = \sum_{\text{pairs } i,j} \mathbf{g}_{ij} \equiv \sum_{i<j} \mathbf{g}_{ij}.$$

For n electrons, there are $n(n-1)/2$ such terms:

Case 1: Diagonal matrix elements ($u = u'$).

$$\langle u\,|\mathbf{A}|\,u\rangle = \int u_1^*(1)\ldots u_N^*(N) \sum_{i<j} \mathbf{g}_{ij} \begin{vmatrix} u_1(1) & \ldots & u_1(N) \\ & \vdots & \\ u_N(1) & \ldots & u_N(N) \end{vmatrix} \mathrm{d}\tau_1 \ldots \mathrm{d}\tau_N$$

Look at *one* of the terms in the sum over i and j, say $i = 1, j = 2$. Then

$$\langle u\,|\mathbf{g}_{12}|\,u\rangle = \int u_1^*(1)\ldots u_N^*(N)\mathbf{g}_{12} \begin{vmatrix} u_1(1) & \ldots & u_1(N) \\ & \vdots & \\ u_N(1) & \ldots & u_N(N) \end{vmatrix} \mathrm{d}\tau_1 \ldots \mathrm{d}\tau_N$$

Of all the terms from the determinant, *two* survive: the principal diagonal and the one which looks like

$$-u_1(2)u_2(1)u_3(3)\ldots u_N(N).$$

The element becomes

$$\langle u | \mathbf{g}_{12} | u \rangle = \int u_1^*(1) u_2^*(2) \mathbf{g}_{12} u_1(1) u_2(2) \, d\tau_1 \, d\tau_2$$

$$- \int u_1^*(1) u_2^*(2) \mathbf{g}_{12} u_1(2) u_2(1) \, d\tau_1 \, d\tau_2$$

Summing over i and j, remembering that the electron indices are dummy, yields

$$\langle u | \sum_{i<j} \mathbf{g}_{ij} | u \rangle = \sum_{i<j} \left[\int u_1^*(i) u_2^*(j) \mathbf{g}_{12} u_1(i) u_2(j) \, d\tau_1 \, d\tau_2 \right.$$

$$\left. - \int u_1^*(i) u_2^*(j) \mathbf{g}_{12} u_1(j) u_2(i) \, d\tau_1 \, d\tau_2 \right]$$

Or

$$\langle u | \mathbf{A} | u \rangle = \sum_{i<j} [\langle i \, j | \mathbf{g}_{12} | i \, j \rangle - \langle i \, j | \mathbf{g}_{12} | j \, i \rangle]$$

Case 2: $u = u'$ except for one spin orbital.

$$\langle u | \sum_{i<j} \mathbf{g}_{ij} | u' \rangle = \sum_{i \neq j} [\langle i \, j | \mathbf{g}_{12} | i' \, j \rangle - \langle i \, j | \mathbf{g}_{12} | j \, i' \rangle]$$

Case 3: $u = u'$ except for two spin orbitals.

$$\langle u | \sum_{i<j} \mathbf{g}_{ij} | u' \rangle = \langle i \, j | \mathbf{g}_{12} | i' \, j' \rangle - \langle i \, j | \mathbf{g}_{12} | j' \, i' \rangle$$

where i' and j' indicate just those spin orbitals of u' which differ from u. All others must be the same, and the spin of i' and j' must be aligned with those of i and j when $\mathbf{g}_{ij}$ is spin independent.

Example: Again choose a $1s2s2p$ configuration, and let

$$| u \rangle = \mathcal{A} | 1 s 0^+ \rangle | 2 s 0^+ \rangle | 2 p 1^+ \rangle$$

Then

$$\langle u | \sum_{i<j} \mathbf{g}_{ij} | u \rangle = \langle (1 s 0^+)_1 (2 s 0^+)_2 | \mathbf{g}_{12} | (1 s 0^+)_1 (2 s 0^+)_2 \rangle$$

$$- \langle (1 s 0^+)_1 (2 s 0^+)_2 | \mathbf{g}_{12} | (1 s 0^+)_2 (2 s 0^+)_1 \rangle$$

$$+ \langle (2 s 0^+)_1 (2 p 1^+)_2 | \mathbf{g}_{12} | (2 s 0^+)_1 (2 p 1^+)_2 \rangle$$

$$- \langle (2 s 0^+)_1 (2 p 1^+)_2 | \mathbf{g}_{12} | (2 s 0^+)_2 (2 p 1^+)_1 \rangle$$

$$+ \langle (1 s 0^+)_1 (2 p 1^+)_2 | \mathbf{g}_{12} | (1 s 0^+)_1 (2 p 1^+)_2 \rangle$$

$$- \langle (1 s 0^+)_1 (2 p 1^+)_2 | \mathbf{g}_{12} | (1 s 0^+)_2 (2 p 1^+)_1 \rangle$$

An expression like $\langle ij|\mathbf{g}_{ij}|ij\rangle$ has not previously been encountered, and an expression for a general two-electron integral can be written as

$$\langle ij|\mathbf{g}_{12}|rt\rangle = \int\int u_1^*(i)u_2^*(j)\mathbf{g}_{12}u_1(r)u_2(t)\,\mathrm{d}\tau_1\,\mathrm{d}\tau_2$$

The diagonal matrix elements of the two-electron operator have special names:

$$\langle ij|\mathbf{g}_{12}|ij\rangle \Rightarrow \text{Coulomb integral}$$

$$\langle ij|\mathbf{g}_{12}|ji\rangle \Rightarrow \text{exchange integral}$$

Antisymmetrization and indistinguishable particles gave us nothing for one-electron operators (or for c-numbers) that would not have resulted by letting $u = u_1(1)u_2(2)\dots u_N(N)$ rather than $u = \mathcal{A}u_1(1)u_2(2)\dots u_N(N)$. The first new feature appears for two-electron operators and is the exchange integral.

Subsequently, explicit formulae for these matrix elements in terms of the $(\mathrm{R}_{n\ell}Y_\ell^m\chi_{m_s})$ one-electron spin orbitals will be presented.

By the way, any problem you are asked to solve will only require the use of diagonal matrix elements. Take note of the fact that what has been derived is valid for both one- and two-electron operators of any functional form. Here our interest is in evaluating the energy, but in subsequent chapters other operators will be of interest.

The relevant energy operators are

$$\mathbf{H} = \sum_i \mathbf{f}_i + \sum_{i<j} \mathbf{g}_{ij} \tag{2.13}$$

$$\mathbf{f}_i \equiv -\tfrac{1}{2}\nabla_i^2 - \frac{Z}{r_i} \tag{2.14}$$

$$\mathbf{g}_{ij} \equiv \frac{1}{r_{ij}} \tag{2.15}$$

Consider now only *diagonal* matrix elements of the Hamiltonian

$$\langle u|\sum_i \mathbf{f}_i|u\rangle = \sum_{\text{shells}} q_{n\ell}\mathrm{I}(n,\ell)$$

$q(n,\ell)$ is the occupation number of a shell (given $n\ell$ value), e.g., for $1s^2 2s 2p^3$, $q_{1s} = 2$, $q_{2s} = 1$, and $q_{2p} = 3$.

$$\mathrm{I}(n,\ell) \equiv \int u^*(n,\ell)\left(-\frac{\nabla^2}{2} - \frac{Z}{r}\right)u(n,\ell)\,\mathrm{d}\tau$$

$$= \int_0^\infty \left[\frac{1}{2}r^{2\ell+2}\frac{\mathrm{d}}{\mathrm{d}r}\left(\frac{\mathrm{R}_{n\ell}^*}{r^\ell}\right)\frac{\mathrm{d}}{\mathrm{d}r}\left(\frac{\mathrm{R}_{n\ell}}{r^\ell}\right) - Z r \mathrm{R}_{n\ell}^*\mathrm{R}_{n\ell}\right]\mathrm{d}r \tag{2.16}$$

Problem 2.8

Prove this.

So $I(n, \ell)$ is the *same* for all states of a given configuration. (The operator $\mathbf{f}_i$ has no off-diagonal elements between states of the same configuration.)

The two-electron operator is more cumbersome:

$$\langle i \, j \mid \frac{1}{r_{ij}} \mid i \, j \rangle = \int \int R^*_{n_i \ell_i}(r_1) R^*_{n_j \ell_j}(r_2) Y^{*m_{\ell_i}}_{\ell_i} Y^{*m_{\ell_j}}_{\ell_j} \left(\frac{1}{r_{12}} \right)$$

$$R_{n_i \ell_i}(r_1) R_{n_j \ell_j}(r_2) Y^{m_{\ell_i}}_{\ell_i}(\theta_1, \phi_1) Y^{m_{\ell_j}}_{\ell_j}(\theta_2, \phi_2) \, d\tau_1 \, d\tau_2 \quad (2.17)$$

The most reasonable way to handle an integral of this form is to invoke the spherical harmonic addition theorem:

$$\frac{1}{r_{12}} = \sum_{k, m_k} \frac{4\pi}{2k+1} \frac{r^k_<}{r^{k+1}_>} Y^{*m_k}_k(\theta_1, \phi_1) Y^{m_k}_k(\theta_2, \phi_2)$$

where

$$\frac{1}{r_{12}} \equiv \frac{1}{|\vec{r}_1 - \vec{r}_2|}$$

and $r_>$ is the larger of $|\vec{r}_1|$ and $|\vec{r}_2|$ and $r_<$ is the smaller of the two (see the appendix *Polynomials and Spherical Harmonics* for the derivation). This is an extremely useful expression but does demand that the user split any integral in which it is invoked into two regions, one in which $|\vec{r}_1|$ is greater than $|\vec{r}_2|$ and one in which the opposite is the case. Reflect on the fact that the angles which originally occurred in the denominator are now in the numerator and are in the form of the spherical harmonics. The price one pays is to have an infinite sum, which at first might seem high, but in fact, the sum is always constrained to the first few terms whenever the operator appears inside a matrix element in which one of the angular momenta is not ridiculously large. The index k cannot be larger than the sum of the angular momenta appearing in the bra and ket making up the matrix element. This is a consequence of what follows.

Our task is to perform the integration expressed in (2.17). There are six integrals in this expression over the variables r_1, r_2, θ_1, θ_2, ϕ_1, and ϕ_2. First look at the integrals just over θ_1 and ϕ_1 for a given value of k and reexpress them without complex conjugates:

$$\left[\frac{4\pi}{2k+1}\right]^{1/2} \sum_{m_k} \int Y_k^{*m_k} Y_{\ell_i}^{*m_{\ell_i}} Y_{\ell_i}^{m_{\ell_i}}(\theta_1, \phi_1)\, d\Omega_1$$

$$= \left[\frac{4\pi}{2k+1}\right]^{1/2} \sum_{m_k} (-1)^{m_k+m_{\ell_i}} \int Y_k^{-m_k} Y_{\ell_i}^{-m_{\ell_i}} Y_{\ell_i}^{m_{\ell_i}}(\theta_1, \phi_1)\, d\Omega_1 \quad (2.18)$$

This integral over three spherical harmonics has a closed form in terms of Clebsch–Gordan coefficients or better still in terms of $3-j$ symbols. Its value will be tabulated in a convenient form, and its derivation would take us too far afield. The result is[2]

$$\int Y_p^{m_p} Y_q^{m_q} Y_s^{m_s}\, d\Omega = \left[\frac{(2p+1)(2q+1)(2s+1)}{4\pi}\right]^{1/2} \begin{pmatrix} p & q & s \\ 0 & 0 & 0 \end{pmatrix} \begin{pmatrix} p & q & s \\ m_p & m_q & m_s \end{pmatrix} \quad (2.19)$$

The last two expressions are $3-j$ symbols. Recall from the defining properties that $p+q+s$ must be an even integer and that $m_p + m_q + m_s = 0$. Applying these to the integral (2.18) results in

$$-m_k = m_p$$
$$-m_{\ell_i} = m_q$$
$$m_{\ell_i} = m_s.$$

So $m_k = 0$. Now the integral can be rewritten as

$$\left[\frac{4\pi}{2k+1}\right]^{1/2} \int Y_k^0 Y_{\ell_i}^{*m_{\ell_i}} Y_{\ell_i}^{m_{\ell_i}}\, d\Omega$$

The coefficient c^k can be defined as

$$c^k(\ell_i m_{\ell_i}; \ell_j m_{\ell_j}) \equiv \left[\frac{4\pi}{2k+1}\right]^{1/2} \int Y_k^{m_{\ell_i}-m_{\ell_j}} Y_{\ell_i}^{*m_{\ell_i}} Y_{\ell_j}^{m_{\ell_j}}\, d\Omega \quad (2.20)$$

[2] A.R. Edmonds *Angular Momentum in Quantum Mechanics* Princeton University Press, Princeton, NJ, 1960.

Shorthand notation for this same integral is $c^k(i; j) \equiv c^k(\ell_i m_{\ell_i}; \ell_j m_{\ell_j})$. The integral above (2.18) is just $c^k(i; i)$. The Coulomb integral that is being considered (2.17) may now be written as

$$\langle i\, j\, | \frac{1}{r_{12}} | i\, j \rangle = \sum_{k=0}^{\infty} c^k(i; i) c^k(j; j) F^k(i, j) \qquad (2.21)$$

where $c^k(i; i)$, the double integral that has just been completed, expresses the integral over θ_1 and ϕ_1. The double integral over θ_2 and ϕ_2 is $c^k(j; j)$ done just as the previous case. What has not yet been done is the double integral over r_1 and r_2 which can be expressed as

$$F^k(i, j) \equiv \int \int R^*_{n_i \ell_i}(r_1) R^*_{n_j \ell_j}(r_2) \frac{r_<^k}{r_>^{k+1}} R_{n_i \ell_i}(r_1) R_{n_j \ell_j}(r_2) r_1^2 r_2^2 \, dr_1 \, dr_2$$

$$(2.22)$$

and $F^k(i, j) = F^k(n_i \ell_i, n_j \ell_j)$. The evaluation of this integral for any specific problem is quite involved and obviously needs the radial wave functions which are specific to each different problem. It suffices for the present discussion to express this integral with notation that is commonly used.

It is customary to define

$$a^k(i; j) \equiv a^k(\ell_i m_{\ell_i}; \ell_j m_{\ell_j}) \equiv c^k(i; i) c^k(j; j)$$

So finally, the combined angular and radial integrals from Eq. (2.17) can be expressed as

$$\langle i\, j\, | \frac{1}{r_{12}} | i\, j \rangle = \sum_{k=0}^{\infty} a^k(i; j) F^k(i, j) \qquad (2.23)$$

In practice, only very few terms are nonzero.

The exchange integral is done in the same manner, but the interchange of i and j on the right side of the ket means that the spin part of the wave functions do not automatically align and if they were not aligned, zero would result. In addition, the radial integral is more complicated than the Coulomb integral:

$$\langle i\, j\, | \frac{1}{r_{12}} | j\, i \rangle = \delta_{m_{s_i} m_{s_j}} \sum_{k=0}^{\infty} \left[c^k(i; j) \right]^2 G^k(i, j) \qquad (2.24)$$

where

$$G^k(i, j) \equiv \int \int R^*_{n_i \ell_i}(r_1) R^*_{n_j \ell_j}(r_2) \frac{r^k_<}{r^{k+1}_>} R_{n_j \ell_j}(r_1) R_{n_i \ell_i}(r_2) r_1^2 r_2^2 \, dr_1 \, dr_2$$

$$(2.25)$$

Problem 2.9

Prove (2.24).

Putting it all together, the diagonal matrix elements of the Hamiltonian between determinantal spin orbitals is given by

$$\langle u | \mathbf{H} | u \rangle = \sum_{\text{shells}} q_{n\ell} I(n, \ell) +$$

$$\sum_{i<j} \sum_{k=0}^{\infty} \left\{ a^k(i; j) F^k(i, j) - \delta_{m_{s_i} m_{s_j}} \left[c^k(i; j) \right]^2 G^k(i, j) \right\}. \quad (2.26)$$

Some remarks are in order. $I(n, \ell)$ will always be a very large negative number. The integrals F^k and G^k are always positive. The energy that results will be negative taken with respect to *complete* ionization of the atom.

Values for c^k and a^k have been tabulated. All that is needed to work out the energy for many excited states of any atom are the values of the Coulomb integrals F^k and the exchange integrals G^k. However, the relative placement of different terms of a configuration can often be done without knowing the values of these integrals.

2.5   Energy Values for Some Simple Examples

Consider the configuration $1s2s$ of helium. There are four states; in the primitive basis of determinantal wave functions, they are

$$u_1 = \mathcal{A} | 1s0^+ \rangle | 2s0^+ \rangle$$

$$u_2 = \mathcal{A} | 1s0^+ \rangle | 2s0^- \rangle$$

$$u_3 = \mathcal{A} | 1s0^- \rangle | 2s0^+ \rangle$$

$$u_4 = \mathcal{A} | 1s0^- \rangle | 2s0^- \rangle$$

In the basis L, S, M_L, and M_S, there are the states

$$^3S_{01} \quad (M_L = 0, M_S = 1)$$

$$^3S_{00}$$

$$^1S_{00}$$

$$^3S_{0-1}$$

Forming a table connecting these basis sets yields

M_L	M_S	m_{ℓ_1}	m_{s_1}	m_{ℓ_2}	m_{s_2}
0	1	0	$1/2$	0	$1/2$
0	0	0	$1/2$	0	$-1/2$
0	0	0	$-1/2$	0	$1/2$
0	-1	0	$-1/2$	0	$-1/2$

Clearly the state $^3S_{01}$ is given by u_1 while $^3S_{0-1}$ is u_4. But $^1S_{00}$ is a linear combination of u_2 and u_3 while $^3S_{00}$ is an orthogonal linear combination of u_2 and u_3.

For the Hamiltonian being considered, all the states of a given *term* are degenerate. The energy of the ^3S term can then be found by

$$E(^3S) = \langle u_1 |\mathbf{H}| u_1 \rangle \quad \text{or} \quad \langle u_4 |\mathbf{H}| u_4 \rangle$$

To obtain the energy of the ^1S, one could proceed in two ways. Evaluate the matrix

$$\langle u_2 |\mathbf{H}| u_2 \rangle \; \langle u_2 |\mathbf{H}| u_3 \rangle$$
$$\langle u_3 |\mathbf{H}| u_2 \rangle \; \langle u_3 |\mathbf{H}| u_3 \rangle$$

and diagonalize it, thereby getting the ^3S and ^1S energies as eigenvalues. There is, however, a much easier way which avoids evaluating off-diagonal matrix elements. Called the Slater sum rule, it reminds you that the trace of a matrix is invariant to a unitary transformation. So the sum of the diagonal matrix elements above must be the sum of the ^3S and ^1S energies. The ^3S energy is known from before, so subtract that from the sum to find the ^1S energy.

Use (2.26) to obtain

$$E(^3S) = \langle u_1 |\mathbf{H}| u_1 \rangle = I(1s) + I(2s) + F^0(1s, 2s) - G^0(1s, 2s)$$

Then

$$\langle u_2 |\mathbf{H}| u_2 \rangle = I(1s) + I(2s) + F^0(1s, 2s)$$

$$\langle u_3 |\mathbf{H}| u_3 \rangle = I(1s) + I(2s) + F^0(1s, 2s)$$

Adding these and subtracting the above gives

$$E(^1S) = I(1s) + I(2s) + F^0(1s,\, 2s) + G^0(1s,\, 2s)$$

So the triplet lies below the singlet. If determinantal wave functions had not been used, one would have obtained $I(1s) + I(2s) + F^0(1s,\, 2s)$ for both of these. The exchange interaction, an electrostatic effect that has nothing to do with any magnetic moment associated with spin, has split the energy between singlet and triplet.

If the configuration had been $1s^22s3s$, the analysis would go through as before, but there would be many more terms to the energy expression:

$$E(1s^22s3s\ ^3S) = 2I(1s) + I(2s) + I(3s)$$
$$+ F^0(1s,\, 1s)$$
$$+ 2F^0(1s,\, 2s) - G^0(1s,\, 2s)$$
$$+ 2F^0(1s,\, 3s) - G^0(1s,\, 3s)$$
$$+ F^0(2s,\, 3s) - G^0(2s,\, 3s)$$
$$E(1s^22s3s\ ^1S) = 2I(1s) + I(2s) + I(3s)$$
$$+ F^0(1s,\, 1s)$$
$$+ 2F^0(1s,\, 2s) - G^0(1s,\, 2s)$$
$$+ 2F^0(1s,\, 3s) - G^0(1s,\, 3s)$$
$$+ F^0(2s,\, 3s) + G^0(2s,\, 3s)$$

Warning: The formulae above are applicable whenever the configuration is $1s^22s3s$, but the actual values of the integrals depend on the value of Z. This configuration exists for Be but also for B^+, C^{++}, N^{+++}, etc., and is called the Be isoelectronic sequence. Spectroscopic notation for these ions is

$$Be\,I, B\,II, C\,III, N\,IV, etc.$$

Consider $2p^2$. Our previous work concluded that there are only three terms: ^3P, ^1S, and ^1D. Refer to the table on page 50. The numbers in the left column label the states in the basis $\left|\, n_i\, \ell_i\, m_{\ell_i}\, m_{s_i}\, \right\rangle$. Use of the Slater sum rule allows us to restrict our attention to just five of these. Using the same numbers as on page 50, there are

			2p		2p	
#	M_L	M_S	m_{ℓ_1}	m_{s_1}	m_{ℓ_2}	m_{s_2}
1	2	0	1	$^1/_2$	1	$-^1/_2$
2	1	1	1	$^1/_2$	0	$^1/_2$
7	0	0	1	$^1/_2$	-1	$-^1/_2$
8	0	0	1	$-^1/_2$	-1	$^1/_2$
9	0	0	0	$^1/_2$	0	$-^1/_2$

Now #1 labels the state $u_1 = \mathcal{A}|2p1^+\rangle|2p1^-\rangle$. In the coupled basis, there is the state $^1D_{20}$ which must equal u_1. The energy of the ^1D can be found by evaluating $\langle u_1 |\mathbf{H}| u_1 \rangle$. Similarly the energy of the ^3P can be found by $\langle u_2 |\mathbf{H}| u_2 \rangle$. Now ^1S has only one state: the $^1S_{00}$. But this must be a linear combination of u_7, u_8, and u_9. The $^1D_{00}$ and $^3P_{00}$ are also orthogonal linear combinations of u_7, u_8, and u_9. So by adding together $\langle u_7 |\mathbf{H}| u_7 \rangle$, $\langle u_8 |\mathbf{H}| u_8 \rangle$, and $\langle u_9 |\mathbf{H}| u_9 \rangle$, one obtains the sum of the ^1D, ^3P and ^1S energies. The first two are known, and by subtraction the energy of the ^1S can be obtained:

$$E(2p^2\, {}^1D) = \langle u_1 |\mathbf{H}| u_1 \rangle$$

(One could just as well use $\langle u_{15} |\mathbf{H}| u_{15} \rangle$.)

$$E(^1D) = 2\mathrm{I}(2p) + \mathrm{F}^0(2p,\, 2p) + \frac{1}{25}\mathrm{F}^2(2p,\, 2p)$$

In like fashion,

$$E(2p^2\, {}^3P) = \langle u_2 |\mathbf{H}| u_2 \rangle$$

(One could just as well use $\langle u_5 |\mathbf{H}| u_5 \rangle$ or $\langle u_6 |\mathbf{H}| u_6 \rangle$ or $\langle u_{10} |\mathbf{H}| u_{10} \rangle$ or $\langle u_{11} |\mathbf{H}| u_{11} \rangle$ or $\langle u_{14} |\mathbf{H}| u_{14} \rangle$.)

$$E(^3P) = 2\mathrm{I}(2p) + \mathrm{F}^0(2p,\, 2p) - \frac{2}{25}\mathrm{F}^2(2p,\, 2p) - \frac{3}{25}\mathrm{G}^2(2p,\, 2p)$$

$$= 2\mathrm{I}(2p) + \mathrm{F}^0(2p,\, 2p) - \frac{1}{5}\mathrm{F}^2(2p,\, 2p)$$

because $\mathrm{G}^k(n\ell,\, n\ell) = \mathrm{F}^k(n\ell,\, n\ell)$.

The desired sum is $\langle u_7 |\mathbf{H}| u_7 \rangle + \langle u_8 |\mathbf{H}| u_8 \rangle + \langle u_9 |\mathbf{H}| u_9 \rangle$.

$$\langle u_7 |\mathbf{H}| u_7 \rangle = 2\mathrm{I}(2p) + \mathrm{F}^0(2p,\, 2p) + \frac{1}{25}\mathrm{F}^2(2p,\, 2p)$$

$$\langle u_8 |\mathbf{H}| u_8 \rangle = 2\mathrm{I}(2p) + \mathrm{F}^0(2p,\, 2p) + \frac{1}{25}\mathrm{F}^2(2p,\, 2p)$$

$$\langle u_9 |\mathbf{H}| u_9 \rangle = 2\mathrm{I}(2p) + \mathrm{F}^0(2p,\, 2p) + \frac{4}{25}\mathrm{F}^2(2p,\, 2p)$$

$$E(^1S) = 2\mathrm{I}(2p) + \mathrm{F}^0(2p,\, 2p) + \frac{2}{5}\mathrm{F}^2(2p,\, 2p)$$

Because the F^k integrals are positive, this analysis tells us that the ^3P lies lowest in energy, the ^1D is next, and the ^1S lies highest. That the ^3P lies lowest is a verification of Hund's rules.

Hund's rules:

1. Of the terms given by equivalent electrons, those with the greatest multiplicity (largest S value) lie deepest, and of these, the lowest is that with the greatest L.

2. Multiplets formed from equivalent electrons are regular when less than half the shell is occupied but inverted when more than half the shell is occupied.

The second rule requires some clarification. When one considers all the levels (nondegenerate for real systems) associated with a given term, that grouping is called a multiplet, e.g., $^3P_{0,1,2}$. Regular spacing means that the level of lowest j lies deepest, while inverted means that the level of highest j lies deepest. The first example in the periodic table of this rule is that the ground state of carbon, $1s^2 2s^2 2p^2$, is $^3P_0$ while the ground state of oxygen, $1s^2 2s^2 2p^4$, is $^3P_2$.

For the configuration $1s^2 2s^2 2p^2$, the energy for the ^1D term is

$$E(1s^2 2s^2 2p^2 \; ^1D) = 2I(1s) + 2I(2s) + 2I(2p)$$
$$+ F^0(1s, \, 1s)$$
$$+ 4F^0(1s, \, 2s) - 2G^0(1s, \, 2s)$$
$$+ 4F^0(1s, \, 2p) - \tfrac{2}{3}G^1(1s, \, 2p)$$
$$+ F^0(2s, \, 2s)$$
$$+ 4F^0(2s, \, 2p) - \tfrac{2}{3}G^1(2s, \, 2p)$$
$$+ F^0(2p, \, 2p) + \tfrac{1}{25}F^2(2p, \, 2p)$$

Every line but the last is the same for each of the three terms ^1D, ^1S, and ^3P. All one needs to obtain the relative placement of the three terms is the value of $F^2(2p, \, 2p)$ for C I:

$$\text{C I}: \quad F^2(2p, \, 2p) = 0.2433 \text{ Hartrees}$$

Problem 2.10

Place the three terms for the lowest configuration of carbon on an energy level diagram with the lowest term at 0 and the top of the diagram at the ionization limit of carbon. Use any consistent set of units.

The energy of C III is a subset of the contributions considered above. Specifically, it has only a single-state $^1S_{00}$ in the ground level whose energy is given by

$$E(\text{C III} \quad 1s^2 2s^2 \; ^1S) = 2I(1s) + 2I(2s)$$
$$+ F^0(1s, \, 1s)$$
$$+ 4F^0(1s, \, 2s) - 2G^0(1s, \, 2s)$$
$$+ F^0(2s, \, 2s)$$

The energy of C I $\quad 1s^2 2s^2 2p^2 \; ^1D$ could then be written as

$$E(\text{C I} \quad 1s^2 2s^2 2p^2\ {}^1D) = E(\text{C III ground state}) + 2I(2p)$$
$$+ 4\text{F}^0(1s,\ 2p) - {}^2\!/\!3\text{G}^1(1s,\ 2p)$$
$$+ 4\text{F}^0(2s,\ 2p) - {}^2\!/\!3\text{G}^1(2s,\ 2p)$$
$$+ \text{F}^0(2p,\ 2p) + {}^1\!/\!25\text{F}^2(2p,\ 2p)$$

If someone were to tell you the energy of C III in the ground state and ask for the energy of some term of C I, your labor has been cut, in this instance, almost in half. For a heavier atom, the saving in effort would be greater still!

2.6 Average Energy of a Configuration

The energy of the three terms ${}^1D$, ${}^1S$, and ${}^3P$ differs only in the coefficients of the $\text{F}^2(2p,\ 2p)$ integral. The energies can be written as

$$E({}^1D) = E_0 + {}^1\!/\!25\text{F}^2(2p,\ 2p)$$
$$E({}^1S) = E_0 + {}^2\!/\!5\text{F}^2(2p,\ 2p) \qquad (2.27)$$
$$E({}^3P) = E_0 - {}^1\!/\!5\text{F}^2(2p,\ 2p)$$

Here E_0 represents all of the energy terms common to the above expressions.

The average energy of the configuration $1s^2 2s^2 2p^2$ is obtained by multiplying each term energy by its statistical weight (the number of states which share that energy), adding them, and dividing by the total number of states, which for this case is 15. The ${}^1D$ has five states, the ${}^3P$ nine, and the ${}^1S$ one. The average energy is then

$$E(av \quad 1s^2 2s^2 2p^2) = E_0 - {}^2\!/\!25\text{F}^2(2p,\ 2p)$$

Each of the term energies could now be written with respect to this average:

$$E({}^1D) = E(av) + {}^3\!/\!25\text{F}^2(2p,\ 2p)$$
$$E({}^1S) = E(av) + {}^{12}\!/\!25\text{F}^2(2p,\ 2p) \qquad (2.28)$$
$$E({}^3P) = E(av) - {}^3\!/\!25\text{F}^2(2p,\ 2p)$$

Expressions for evaluating the average energy of the configuration have been derived in Condon and Odabaşi.[3] Such expressions might save a significant amount of time for performing calculations, by hand, for atoms heavier than neon. Practically

---

[3]*Atomic Structure*, E.U. Condon and Halis Odabaşi, Cambridge University Press, 1980.

speaking, one rarely wishes to do that, and the expressions given here enable one to understand what the computer is doing when such evaluations are performed numerically.

2.7 Hartree–Fock Equations

This section will offer a very brief look at the Hartree–Fock equations written with a notation different from what has been used previously in this book.[4] The reason for doing this is that one sees many different sets of notations and it is useful to learn their equivalence. Let $\mathbf{H} = \mathbf{H}_1 + \mathbf{H}_2$:

$$\mathbf{H}_1 \equiv \sum_{i=1}^{N} \mathbf{f}_i = \sum \left[-\frac{\nabla_i^2}{2} - \frac{Z}{r_i} \right]$$

$$\mathbf{H}_2 = \sum_{i<j=1}^{N} \frac{1}{r_{ij}}; \qquad r_{ij} \equiv |\vec{r}_i - \vec{r}_j|$$

An antisymmetrized product of one-electron spin orbitals will form the basis and be written as

$$\Phi \equiv \mathcal{A} u_1(q_1) u_2(q_2) \ldots u_N(q_N)$$

The total energy can be written as (compare to (2.26))

$$E[\Phi] = \sum_{\lambda} I_\lambda + \tfrac{1}{2} \sum_{\lambda} \sum_{\mu} \left[J_{\lambda\mu} - \delta_{m_{s_\lambda} m_{s_\mu}} K_{\lambda\mu} \right]$$

$$I_\lambda \equiv \langle\, u_\lambda(q_i) \, |\mathbf{f}_i| \, u_\lambda(q_i) \,\rangle$$

$$J_{\lambda\mu} \equiv \langle\, u_\lambda(q_i) u_\mu(q_j) \, | \frac{1}{r_{ij}} | \, u_\lambda(q_i) u_\mu(q_j) \,\rangle$$

$$K_{\lambda\mu} \equiv \langle\, u_\lambda(q_i) u_\mu(q_j) \, | \frac{1}{r_{ij}} | \, u_\mu(q_i) u_\lambda(q_j) \,\rangle$$

and λ, μ stand for a set of quantum numbers

$$\lambda \Longleftrightarrow n_\lambda \ell_\lambda m_{\ell_\lambda} m_{s_\lambda}$$

---

[4]Robert D. Cowan *Theory of Atomic Structure and Spectra* University of California Press, 1981.

and i and j serve to distinguish two different sets of electron coordinates, but *any of the two*—there is no sum over i and j. To try to relate this notation to what was done previously, note that $J_{\lambda\mu}$ is given by (2.23) while $K_{\lambda\mu}$ is given by (2.24).

Using the method of Lagrange multipliers a variational equation can be written as

$$\delta E - \sum_{\lambda,\mu} \epsilon_{\lambda\mu} \delta \langle u_\mu \,|\, u_\lambda \rangle = 0$$

By performing a unitary transformation the above equation can be written in diagonal form as

$$\delta E - \sum_\lambda \epsilon_\lambda \delta \langle u_\lambda \,|\, u_\lambda \rangle = 0$$

Performing the variation on the expression for the total energy yields the Hartree–Fock equations for closed shells:

$$\left[-\frac{\nabla_i^2}{2} - \frac{Z}{r_i} \right] u_\lambda(q_i) + \left[\sum_\mu \int u_\mu^*(q_j) \frac{1}{r_{ij}} u_\mu(q_j)\, dq_j \right] u_\lambda(q_i)$$

$$- \sum_\mu \left[\delta_{m_{s_\lambda} m_{s_\mu}} \int u_\mu^*(q_j) \frac{1}{r_{ij}} u_\lambda(q_j)\, dq_j \right] u_\mu(q_i) = \epsilon_\lambda u_\lambda(q_i)$$

These are a set of coupled integrodifferential equations which have replaced the Schrödinger equation for a multi-electron atom. Indistinguishability of particles, antisymmetrization (Pauli principle), and the form of one-electron spin orbitals (appropriate for a central potential) have all been built in.

ϵ_λ is approximately the energy needed to remove the electron with quantum numbers λ. This is Koopman's theorem.

The equations are solved iteratively for the functions $u_\lambda(q_i)$. Note that the noncentral potential, $\mathbf{H}_2$, has been integrated over, forming an effective central potential. This is the potential alluded to early in this chapter when describing the self-consistent field approximation.

2.8 Tables

Table 2.1 $a^k(\ell_i m_{\ell_i}; \ell_j m_{\ell_j})$

	m_{ℓ_i}	m_{ℓ_j}	k		
			0	2	4
ss	0	0	1	0	0
sp	0	±1	1	0	0
	0	0	1	0	0
sd	0	±2	1	0	0
	0	±1	1	0	0
	0	0	1	0	0
pp	±1	±1	1	1/25	0
	±1	0	1	−2/25	0
	0	0	1	4/25	0
pd	±1	±2	1	2/35	0
	±1	±1	1	−1/35	0
	±1	0	1	−2/35	0
	0	±2	1	−4/35	0
	0	±1	1	2/35	0
	0	0	1	4/35	0
dd	±2	±2	1	4/49	1/441
	±2	±1	1	−2/49	−4/441
	±2	0	1	−4/49	6/441
	±1	±1	1	1/49	16/441
	±1	0	1	2/49	−24/441
	0	0	1	4/49	36/441

Table 2.2 $c^k(\ell_i m_{\ell_i}; \ell_j m_{\ell_j})$

	m_{ℓ_i}	m_{ℓ_j}	k 0	1	2	3	4
ss	0	0	1	0	0	0	0
sp	0	± 1	0	$-\sqrt{1/3}$	0	0	0
	0	0	0	$\sqrt{1/3}$	0	0	0
sd	0	± 2	0	0	$\sqrt{1/5}$	0	0
	0	± 1	0	0	$-\sqrt{1/5}$	0	0
	0	0	0	0	$\sqrt{1/5}$	0	0
pp	± 1	± 1	1	0	$-\sqrt{1/25}$	0	0
	± 1	0	0	0	$\sqrt{3/25}$	0	0
	± 1	∓ 1	0	0	$-\sqrt{6/25}$	0	0
	0	0	1	0	$\sqrt{4/25}$	0	0
pd	± 1	± 2	0	$-\sqrt{6/15}$	0	$\sqrt{3/245}$	0
	± 1	± 1	0	$\sqrt{3/15}$	0	$-\sqrt{9/245}$	0
	± 1	0	0	$-\sqrt{1/15}$	0	$\sqrt{18/245}$	0
	± 1	∓ 1	0	0	0	$-\sqrt{30/245}$	0
	± 1	∓ 2	0	0	0	$\sqrt{45/245}$	0
	0	± 2	0	0	0	$\sqrt{15/245}$	0
	0	± 1	0	$-\sqrt{3/15}$	0	$-\sqrt{24/245}$	0
	0	0	0	$\sqrt{4/15}$	0	$\sqrt{27/245}$	0
dd	± 2	± 2	1	0	$-\sqrt{4/49}$	0	$\sqrt{1/441}$
	± 2	± 1	0	0	$\sqrt{6/49}$	0	$-\sqrt{5/441}$
	± 2	0	0	0	$-\sqrt{4/49}$	0	$\sqrt{15/441}$
	± 2	∓ 1	0	0	0	0	$-\sqrt{35/441}$
	± 2	∓ 2	0	0	0	0	$\sqrt{70/441}$
	± 1	± 1	1	0	$\sqrt{1/49}$	0	$-\sqrt{16/441}$
	± 1	0	0	0	$\sqrt{1/49}$	0	$\sqrt{30/441}$
	± 1	∓ 1	0	0	$-\sqrt{6/49}$	0	$-\sqrt{40/441}$
	0	0	1	0	$\sqrt{4/49}$	0	$\sqrt{36/441}$

Table 2.3 Russell–Saunders coupling for equivalent electrons[a]

Config.	Terms
s	^2S
p, p^5	^2P
p^2, p^4	$^1S\,D$ ^3P
p^3	$^2P\,D$ ^4S
d, d^9	^2D
d^2, d^8	$^1S\,D\,G$ $^3P\,F$
d^3, d^7	$^2P\,D\,F\,G\,H$ (D: 2) $^4P\,F$
d^4, d^6	$^1S\,D\,F\,G\,I$ (2 2 2) $^3P\,D\,F\,G\,H$ (2 2) ^5D
d^5	$^2S\,P\,D\,F\,G\,H\,I$ (3 2 2) $^4P\,D\,F\,G$ ^6S
f, f^{13}	^2F
f^2, f^{12}	$^1S\,D\,G\,I$ $^3P\,F\,H$
f^3, f^{11}	$^2P\,D\,F\,G\,H\,I\,K\,L$ (2 2 2 2) $^4S\,D\,F\,G\,I$
f^4, f^{10}	$^1S\,D\,F\,G\,H\,I\,K\,L\,N$ (2 4 4 2 3 2) $^3P\,D\,F\,G\,H\,I\,K\,L\,M$ (3 2 4 3 4 2 2) $^5S\,D\,F\,G\,I$
f^5, f^9	$^2P\,D\,F\,G\,H\,I\,K\,L\,M\,N\,O$ (4 5 7 6 7 5 5 3 2) $^4P\,D\,F\,G\,H\,I\,K\,L\,M$ (2 3 4 4 3 3 2) $^6P\,F\,H$
f^6, f^8	$^1S\,P\,D\,F\,G\,H\,I\,K\,L\,M\,N\,Q$ (4 6 4 8 4 7 3 4 2 2) $^3P\,D\,F\,G\,H\,I\,K\,L\,M\,N\,O$ (6 5 9 7 9 6 6 3 3) $^5S\,P\,D\,F\,G\,H\,I\,K\,L$ (3 2 3 2 2) ^7F
f^7	$^2S\,P\,D\,F\,G\,H\,I\,K\,L\,M\,N\,O\,Q$ (2 5 7 10 10 9 9 7 5 4 2) $^4S\,P\,D\,F\,G\,H\,I\,K\,L\,M\,N$ (2 2 6 5 7 5 5 3 3) $^6P\,D\,F\,G\,H\,I$ ^8S

[a]The number of terms of the given type is given under the term designation

Chapter 3
Electro- and Magnetostatic Interactions

This chapter will include all of those static interactions that can in some way effect the total energy of any given level of an atom or ion of interest. That's a tall order. Consider that the total energy of an atom is known, having solved the problem presented in the previous chapter; that is, all of the electrons are in stationary states about some nucleus, with the only interactions being electrostatic attraction to the nucleus and repulsion to other electrons. The solution to such a problem would have ignored additional interactions. The largest, and one alluded to in previous chapters, is the fine-structure interaction. This is the coupling of the angular momenta of the outer electrons, the ones in open subshells. All filled shells and subshells have zero total angular momentum, so those electrons cannot take part. Whichever way one chooses to couple the angular momenta, and there are many different possible ways (Russell–Saunders coupling being just one), the total angular momentum, J, is a good quantum number.

At this stage the question you might want to ask is why aren't all of the different levels with differing J values degenerate? The answer is that if the Hamiltonian were composed only of the terms considered in the previous chapter, they all would be degenerate. But measurement shows that they are not, so something must have been left out of the Hamiltonian. Actually quite a few terms have been left out of the Hamiltonian which is known by solving the one-electron hydrogen atom relativistically. In recent years, the two-electron atom has been solved relativistically (not in closed form but to extraordinary precision), so it really is known that there are lots of interactions among the orbital and spin angular momenta of the electrons that produce additional terms in the Hamiltonian. The good news is that they can be handled quite successfully by perturbation techniques.

The plot thickens when one considers the fact that the nucleus also can have a net angular momentum with an associated magnetic dipole moment, and an electric quadrupole moment and these too can interact with the electron cloud to alter the total energy and produce a new total angular momentum. What is interesting is that these nuclear effects are typically orders of magnitude smaller than the fine-structure effects mentioned above, and if one is performing an experiment to a

R.L. Brooks, *The Fundamentals of Atomic and Molecular Physics*, Undergraduate Lecture Notes in Physics, DOI 10.1007/978-1-4614-6678-9_3,
© Springer Science+Business Media New York 2013

level of precision that is not sensitive to such small interactions, then ignoring those interactions is OK. We do, however, live in a world of high-precision measurements, so it is commonly the case that one has to consider the specific atom and levels of interest before deciding on which interactions to include.[1]

The above comments still ignore external electric and magnetic fields. Interactions with electric fields are called the Stark effect, while those with magnetic fields are called the Zeeman effect. For typical laboratory fields the effects on energy from these interactions are intermediate between fine structure and hyperfine structure. But both of these effects might be relevant to an atom you care to investigate even when you don't deliberately apply external fields. The earth's magnetic field can easily perturb a high-precision measurement. An ion moving in an ion beam experiences a motional electric field by moving in the earth's magnetic field. An ion in a crystal experiences an electric field from the surrounding crystal lattice. As mentioned above, the electric and magnetic fields of the nucleus interact with the electron cloud. Hence, the effects of external fields will be taken up after fine structure and before hyperfine structure.

Modern experimental atomic physics offers a large selection of interactions that can perturb an atom or ion of interest. What we want to strive for is the basic understanding of how quantum mechanics can be applied to a multi-electron atom subject to the interactions just mentioned. These are the interactions that will comprise this chapter.

3.1   Fine Structure

The largest term missing from the Hamiltonian of Chap. 2 is the fine-structure contribution, which for a multi-electron atom can be written as

$$\mathbf{H}_{FS} = \frac{Z\alpha^2}{2} \sum_{i=1}^{N} f(r_i)\vec{\ell}_i \cdot \vec{s}_i \qquad (3.1)$$

where the sum is over the N electrons in the atom. For hydrogen, previously, the additional term was written $f(r) = 1/r^3$, and this is the form commonly taken for f. However, self-consistent field treatments can use a somewhat more complicated expression. Pause for a moment to reflect on the fact that this operator, as written, is a simple sum over all electrons of the operator that was introduced in Chap. 1. One can think of it as an interaction between the electron's magnetic moment and

---

[1]It seems appropriate at this point to introduce the reader to an on-line service that lists basic atomic spectroscopic data including the lower-lying energy levels of most atoms and ions. The American National Institute of Standards and Technology (NIST) maintains the Web site http:// physics.nist.gov/PhysRefData/Handbook/index.html from which it is straightforward to obtain basic spectroscopic data.

the magnetic field caused by the orbital motion of a charged particle, the very same electron. But we are now considering a multi-electron problem, so how can we know in advance that summing this one-electron operator captures all of the relevant interactions? In particular, when looking at the relativistic solution of the helium atom, there occur terms involving the interaction of an electron's spin with the spin of another electron as well as the interaction of an electron's spin with the orbital motion of another electron. So, yes, there are terms in the Hamiltonian being left out. Are these bigger or smaller than the one being considered? Interestingly, for helium, they are of comparable size. For light atoms heavier than helium, maybe up to krypton, this spin-orbit interaction is significantly larger than the others and represents the one most commonly considered.

Let us rewrite the defining equation above as

$$\mathbf{H}_{FS} = \sum_i \zeta_i(r_i)\vec{\ell}_i \cdot \vec{s}_i \tag{3.2}$$

and not concern ourselves about calculating ζ_i. It is often evaluated empirically. It is important to remember that $\zeta_i(r_i)$ is a one-electron operator, and as such, when its value in a matrix element is used, only quantum number designators are needed and it is written as $\zeta_{n\ell}$ or more completely as $\zeta(n_i\ell_i)$.

Pause for a moment and reflect on the fact that the Hamiltonian of Eq. (3.2) does not use the simpler operator $\vec{\mathbf{L}} \cdot \vec{\mathbf{S}}$ which is commonly used in elementary treatments of the fine structure. This simpler operator will arise naturally when the treatment of the more complete operator of Eq. (3.2) is taken up.

To obtain any corrections to the term energies and to obtain the level energies, allowing for the lifting of any degeneracies that might occur, the technique of *direct diagonalization* will be used which is usually taught as degenerate perturbation theory.

In the basis $LSJM_J$ one needs to examine matrix elements of the form

$$\langle \alpha\, L\, S\, J\, M_J | \mathbf{H}_{FS} | \alpha\, L'\, S'\, J'\, M_J' \rangle$$

One usually chooses a manifold of states which lie reasonably close together such as those of a *given configuration*; hence, α is the same on both the bra and ket of the matrix element above. Evaluating such matrix elements is hopelessly complicated in all but the simplest cases. The operator is *not* diagonal in this basis or in any other basis that has been considered so far.

In order to grasp the levels of approximation that are usually applied, let us consider a specific example, such as the $2p^2$ configuration, which has ^3P, ^1S, and ^1D terms.

Recall that a ket such as $|LSJM_J\rangle$ must be converted to a ket such as $|LSM_LM_S\rangle$ via Clebsch–Gordan coefficients. This must then be decomposed into the primitive one-electron basis using ladder operators or projection techniques. It is not useful for you to know *how* to do that, but rather that it has to be done. The kets that diagonalize $\mathbf{H}_0$, the Hamiltonian of our previous chapter, can be expressed as

$$M_J = 2 \quad {}^3P_2 : (1^+0^+)$$

$$ \quad {}^1D_2 : (1^+1^-)$$

$$M_J = 1 \quad {}^3P_2 : \frac{1}{2}\left[(1^+0^-) + (1^-0^+)\right] + \frac{1}{\sqrt{2}}(1^+\text{-}1^+)$$

$$ \quad {}^3P_1 : \frac{1}{2}\left[(1^+0^-) + (1^-0^+)\right] - \frac{1}{\sqrt{2}}(1^+\text{-}1^+)$$

$$ \quad {}^1D_2 : \frac{1}{\sqrt{2}}\left[(1^+0^-) - (1^-0^+)\right]$$

$$M_J = 0 \quad {}^3P_2 : \frac{1}{\sqrt{6}}\left[(1^-0^-) + (0^+\text{-}1^+)\right] + \frac{1}{\sqrt{3}}\left[(1^+\text{-}1^-) + (1^-\text{-}1^+)\right]$$

$$ \quad {}^3P_1 : \frac{1}{\sqrt{2}}\left[(1^-0^-) - (0^+\text{-}1^+)\right]$$

$$ \quad {}^3P_0 : \frac{1}{\sqrt{3}}\left[(1^-0^-) + (0^+\text{-}1^+)\right] - \frac{1}{\sqrt{6}}\left[(1^+\text{-}1^-) + (1^-\text{-}1^+)\right]$$

$$ \quad {}^1D_2 : \frac{1}{\sqrt{6}}\left[(1^+\text{-}1^-) - (1^-\text{-}1^+) + 2(0^+0^-)\right]$$

$$ \quad {}^1S_0 : \frac{1}{\sqrt{3}}\left[(1^+\text{-}1^-) - (1^-\text{-}1^+) - (0^+0^-)\right] \tag{3.3}$$

The complete matrix of $\mathbf{H}_{FS}$ in the $LSJM_J$ basis is 15×15. However, each value of M_J yields the same result because of the isotropy of space. Furthermore the rest of the Hamiltonian is diagonal in this basis. Remember that the three relevant terms differ in energy only by factors of F^2 and that the rest of the diagonal energy, E_0, should be added to every term on the diagonal. It is absent only to make the matrix more readable. The relevant energy expression for $\mathbf{H}_0$ is given by the set of equations (2.27). The energy expression for the complete Hamiltonian is then given by the matrix below:

$$\langle\, 2p^2\, L\, S\, J\, |\mathbf{H}_0 + \mathbf{H}_{FS}|\, 2p^2\, L'\, S'\, J'\,\rangle = \tag{3.4}$$

	${}^1D_2$	${}^3P_2$	${}^3P_1$	${}^3P_0$	${}^1S_0$
${}^1D_2$	$F^2/25$	$\zeta/\sqrt{2}$	0	0	0
${}^3P_2$	$\zeta/\sqrt{2}$	$-F^2/5 + \zeta/2$	0	0	0
${}^3P_1$	0	0	$-F^2/5 - \zeta/2$	0	0
${}^3P_0$	0	0	0	$-F^2/5 - \zeta$	$-\sqrt{2}\zeta$
${}^1S_0$	0	0	0	$-\sqrt{2}\zeta$	$2F^2/5$

To see how one gets such a result, consider the matrix element

$$\langle\, {}^1D_2\, |\mathbf{H}_{FS}|\, {}^3P_2\,\rangle = \langle\, {}^1D_{22}\, |\mathbf{H}_{FS}|\, {}^3P_{22}\,\rangle$$

Since any M_J could be chosen (remember the degeneracy) the simplest one uses $M_J = 2$. The fine-structure operator for two p electrons may be written as

$$\mathbf{H}_{FS} = \zeta(2p)\vec{\ell}_1 \cdot \vec{s}_1 + \zeta(2p)\vec{\ell}_2 \cdot \vec{s}_2 \qquad (3.5)$$

But $\vec{\ell}_1 \cdot \vec{s}_1 = \frac{1}{2}(\ell_1^+ s_1^- + \ell_1^- s_1^+) + \ell_1^z s_1^z$ where $\ell^+ \equiv \ell_x + i\ell_y$.

Problem 3.1

Show that $\vec{\ell} \cdot \vec{s} = \frac{1}{2}(\ell^+ s^- + \ell^- s^+) + \ell^z s^z$.

It follows that

$$\mathbf{H}_{FS} = \frac{\zeta}{2}\left(\ell_1^+ s_1^- + \ell_1^- s_1^+\right) + \zeta\ell_1^z s_1^z + \frac{\zeta}{2}\left(\ell_2^+ s_2^- + \ell_2^- s_2^+\right) + \zeta\ell_2^z s_2^z$$

(3.4) becomes

$$\langle 1^+ 1^- |\zeta(\vec{\ell}_1 \cdot \vec{s}_1 + \vec{\ell}_2 \cdot \vec{s}_2)| 1^+ 0^+ \rangle$$

The operator must convert $1^+ 0^+$ into $1^+ 1^-$ or else othogonality yields zero. Only $\ell_2^+ s_2^-$ does that. Recall

$$\ell_2^+ | 1^+ 0^+ \rangle = \ell_2^+ | 1 \overset{m_\ell \, m_s}{\tfrac{1}{2}} \rangle_1 | 0 \overset{m_\ell \, m_s}{\tfrac{1}{2}} \rangle_2.$$

The general expression for the raising operator is $j^+ | j\, m \rangle = [(j - m)(j + m + 1)]^{1/2} | j\, m + 1 \rangle$. For this case, $j = \ell = 1$ and $m = m_\ell = 0$.

$$\ell_2^+ | 1^+ 0^+ \rangle = \sqrt{2} | 1^+ 1^+ \rangle$$

$$s_2^- \left\{ \sqrt{2} | 1^+ 1^+ \rangle \right\} = \sqrt{2} | 1^+ 1^- \rangle$$

So

$$\langle {}^1D_2 |\mathbf{H}_{FS}| {}^3P_2 \rangle = \frac{\sqrt{2}}{2}\zeta = \frac{\zeta}{\sqrt{2}}$$

The diagonal matrix elements are given by

$$\langle {}^1D_2 |\mathbf{H}_{FS}| {}^1D_2 \rangle = \langle 1^+ 1^- |\mathbf{H}_{FS}| 1^+ 1^- \rangle$$

The ket on the right must not be raised or lowered, so only the z part of the operator applies, and

$$\langle 1^+ 1^- |\zeta(\ell_1^z s_1^z + \ell_2^z s_2^z)| 1^+ 1^- \rangle = 0$$

Finally,

$$\langle {}^3P_2 |\mathbf{H}_{FS}| {}^3P_2 \rangle = \langle 1^+ 0^+ |\zeta(\ell_1^z s_1^z + \ell_2^z s_2^z)| 1^+ 0^+ \rangle = \frac{\zeta}{2}$$

Problem 3.2

Work out the remaining matrix elements of $\mathbf{H}_{FS}$.

Problem 3.3

If one diagonalizes the Hamiltonian under the approximation that $\zeta \ll F^2 < E_0$, show that the diagonal entries are perturbed by a term of order ζ^2/F^2.

The conditions specified in the previous problem are those required for LS coupling to be appropriate. When these conditions hold, to a good approximation *the diagonal entries alone* are the energies. With E_0 the same as given by Eq. (2.27), this yields

$$E(^1S_0) = E_0 + \tfrac{2}{5}F^2$$

$$E(^1D_2) = E_0 + \tfrac{1}{25}F^2$$

$$E(^3P_2) = E_0 - \tfrac{1}{5}F^2 + \zeta/2 \tag{3.6}$$

$$E(^3P_1) = E_0 - \tfrac{1}{5}F^2 - \zeta/2$$

$$E(^3P_0) = E_0 - \tfrac{1}{5}F^2 - \zeta$$

This represents the Landé approximation in which

$$\mathbf{H}_{FS} = \sum_i \zeta_i \vec{\ell}_i \cdot \vec{s}_i \rightarrow \Gamma \vec{\mathbf{L}} \cdot \vec{\mathbf{S}}$$

Then

$$\boxed{\langle \alpha\, L\, S\, J\, |\mathbf{H}_{FS}| \alpha\, L\, S\, J \rangle = \frac{\Gamma}{2}\, [J(J+1) - L(L+1) - S(S+1)]} \tag{3.7}$$

From our example above $\Gamma = \zeta/2$. So the elementary treatment of fine structure results by ignoring off-diagonal matrix elements. When ζ is large (compared to F^2), these terms cannot be ignored and the matrix must be diagonalized. This situation is called intermediate coupling. For two p electrons outside of a closed shell the above approximation works well throughout the *carbon* column of the periodic table until one reaches *lead*, atomic number 82. For this element, the energy values listed by NIST[2] afford an opportunity to show that the method of direct diagonalization as outlined here works quite well. When solving problems, don't let the zero of energy be a concern. Atomic energy-level tables take the zero to be the lowest energy level

---

[2]The American National Institute of Standards and Technology (NIST); see http://physics.nist.gov/cgi-bin/ASD/energy1.pl.

Table 3.1 Energy levels of
neutral silicon

Configuration	Term	J	Level (cm^{-1})
$3s^2 3p^2$	^3P	0	0.000
		1	77.112
		2	223.157
	^1D	2	6,298.847
	^1S	0	15,394.362

of the atom or ion of interest. Simply subtract the lowest-lying energy from every
calculated level when making comparisons.

Problem 3.4

Using direct diagonalization with $\zeta_{6p} = 7{,}280$ cm^{-1} and $F^2 = 23{,}200$ cm^{-1}, show that the five
lowest-lying energy levels for Pb can be obtained to better than 2 %.

Even within the Landé approximation, Γ is a function of the individual ζ_i and
moreover is also a function of the orbital quantum numbers. Γ is different for every
term of a configuration, so it is written as $\Gamma(LS)$.

For two electrons outside of a closed shell, one can derive the following useful
expression for Γ:

$$\Gamma(LS) = \zeta(n_1 \ell_1) \left[\frac{L(L+1) + \ell_1(\ell_1+1) - \ell_2(\ell_2+1)}{4L(L+1)} \right]$$
$$+ \zeta(n_2 \ell_2) \left[\frac{L(L+1) + \ell_2(\ell_2+1) - \ell_1(\ell_1+1)}{4L(L+1)} \right]. \quad (3.8)$$

From (3.7) it can be seen that two energy levels of a given term (same S and
L) having $J_{\text{upper}} = J$ and $J_{\text{lower}} = J - 1$ are separated by ΓJ. This is the Landé
interval rule, useful for practical spectroscopy. Of course, this rule applies only when
LS coupling is good.

This is perhaps a good place to mention that helium, a light element for which
LS coupling is good, does not satisfy the Landé interval rule. The reason is that
the fine-structure interaction of Eq. (3.2) is the first in a series of interaction terms.
For helium, the higher-order terms are of comparable magnitude to the term already
considered, whereas for higher Z, this first-order term dominates. Then the warning
is don't try to apply these expressions to helium.

Staying within the np^2 terms, *silicon* affords a nice example of the Landé interval
rule. Consider the energies as written in the NIST table (Table 3.1).

The energy interval between $J = 2$ and $J = 1$ of 146 cm^{-1} is very nearly twice
the interval from $J = 1$ to $J = 0$ of 77 cm^{-1}.

The Landé formula, (3.7), yields good results whenever the magnitude of $\zeta_{n\ell}$ is small compared to the separation of term energies. For the silicon example above, the formula will give the splitting among the $^{3}P_{2,1,0}$ levels so long as the $^{1}D_{2}$ and $^{1}S_{0}$ levels are far away, which they are. It's a matter of judgment regarding what "far away" means but certainly an order of magnitude should be sufficient. When the energy separation between terms is comparable to or not much larger than the fine-structure splitting within one term, there is an interaction between all levels with the same J value.

Two further points, not easy to prove, are worth mentioning:

1. For k electrons in the same subshell, which when filled holds m electrons, one may write

$$\Gamma(n\ell^{m-k} \, LS) = -\Gamma(n\ell^{k} \, LS)$$

2. As a corollary to the above, $\Gamma(LS)$ of half-filled shells is zero.

As you have probably realized by now, the name "fine structure" can be a misnomer for heavier atoms and ions. The interaction scales as Z^4 so that it is really only "fine" for the light elements. It ranges from a fraction of a cm^{-1} to several thousand cm^{-1}.

Problem 3.5

Why does the fine-structure interaction scale as Z^4?

3.2 Zeeman Effect

The Zeeman effect is the name given to the splitting observed in spectral lines when the source of those lines is placed in an external magnetic field. The quantum mechanical explanation, which will be developed here, rests on two important facts. The first is that a single spectral line, representing the transition energy between two levels, is not really a single line at all but is composed of a number of lines all occurring at the same energy. This is a consequence of the degeneracy of most energy levels related to the isotropy of space and described by the m quantum number. The second is that there is an interaction energy between any magnetic moment that the atom of interest may possess and the applied external magnetic field. If the total angular momentum of the electron cloud were zero, there would be no magnetic moment and no interaction energy for that level. Both the upper and lower levels would have to have zero total angular momentum for there to be no observable Zeeman effect for that transition. For reasons discussed in the following chapter, such transitions are highly forbidden.

A reasonably complete derivation of the interaction energy of a classical magnetic dipole placed in an external magnetic field is given in an appendix. The magnetic dipole of an orbiting charged particle is elementary, and the expression for a collection of charges having the same charge to mass ratio is given by

$$\vec{\mu}_\ell = -\sum_{i=1}^{N} \frac{1}{2} \left(\frac{e}{mc}\right) \vec{\ell}_i. \tag{3.9}$$

The quantum mechanical validity of this expression will be assumed where $\vec{\mu}_\ell$ and $\vec{\ell}_i$ are considered to be operators. In quantum mechanics the spin of the electron is also an angular momentum and one might suppose that there would be a magnetic moment associated with it. There is, and the value is given by

$$\vec{\mu}_s = -\sum_{i=1}^{N} \left(\frac{e}{mc}\right) \vec{s}_i \tag{3.10}$$

Note that this is a factor of two larger than might have been guessed. It is important to realize that no value for this coefficient should come as a surprise since no attempt has been made to derive or justify any value whatever. In fact, the best value is somewhat larger than 2. The number 2 comes out of Dirac theory, while the deviation from 2 is a quantum electrodynamic effect.

The total magnetic moment is given by

$$\vec{\mu} = \vec{\mu}_\ell + \vec{\mu}_s = -\frac{1}{2} \left(\frac{e}{mc}\right) (\vec{L} + 2\vec{S}) \tag{3.11}$$

or more properly as

$$\vec{\mu} = \vec{\mu}_\ell + \vec{\mu}_s = -\frac{1}{2} \left(\frac{e}{mc}\right) (\vec{L} + g\vec{S})$$

where g is called the magnetic g-factor of the electron. Measurements of g are among the most precise in all of physics. For our work assume a value of $g = 2$. Equation (3.11) can also be written as

$$\vec{\mu} = -\frac{1}{2} \left(\frac{e}{mc}\right) (\vec{J} + \vec{S})$$

Writing $\vec{J}$ and $\vec{S}$ without dimensions gives

$$\vec{\mu} = -\frac{1}{2} \frac{e\hbar}{mc} (\vec{J} + \vec{S}) = -\mu_0 (\vec{J} + \vec{S}) \tag{3.12}$$

where[3]

$$\mu_0 = \text{Bohr magneton} \equiv \frac{1}{2}\frac{e\hbar}{mc} = 9.274078 \cdot 10^{-24}\ \frac{\text{J}}{\text{Tesla}} = 9.274078 \cdot 10^{-21}\ \frac{\text{erg}}{\text{gauss}}$$

Then the interaction energy is

$$W = -\vec{\mu} \cdot \vec{B} = \mu_0(\mathbf{\vec{J}} + \mathbf{\vec{S}}) \cdot \vec{B} \tag{3.13}$$

Since an external field selects a spatial direction, no generality is lost by taking the axis of quantization (z) to lie along the $\vec{B}$ field. Then

$$W = \mu_0(\mathbf{J}_z + \mathbf{S}_z)B \tag{3.14}$$

In the basis $LSJM_J$, $\mathbf{S}_z$ is a noncommuting operator. Of course one could work out the matrix elements in the basis LSM_LM_S, but in that basis the fine-structure interaction $\alpha\vec{\mathbf{L}} \cdot \vec{\mathbf{S}}$ is non-diagonal.

Prior to the development of quantum mechanics, Landé worked out a formula based on the vector model in which he replaced (3.13) by the relation

$$W \approx \mu_0 \frac{\left[(\mathbf{\vec{J}} + \mathbf{\vec{S}}) \cdot \mathbf{\vec{J}}\right](\mathbf{\vec{J}} \cdot \vec{B})}{|\mathbf{\vec{J}}|^2}$$

Since $\mathbf{\vec{J}} \cdot \mathbf{\vec{S}} = \tfrac{1}{2}(|\mathbf{\vec{J}}|^2 + |\mathbf{\vec{S}}|^2 - |\mathbf{\vec{L}}|^2)$ and $\mathbf{\vec{J}} \cdot \vec{B} = \mathbf{J}_z B$, the above becomes

$$W = \mu_0 B \frac{\left[|\mathbf{\vec{J}}|^2 + \tfrac{1}{2}(|\mathbf{\vec{J}}|^2 + |\mathbf{\vec{S}}|^2 - |\mathbf{\vec{L}}|^2)\right]}{|\mathbf{\vec{J}}|^2}\mathbf{J}_z$$

Now the change in the energy level is given by

$$\Delta E = \langle \alpha L S J M_J | W | \alpha L S J M_J \rangle$$

which is readily worked out to be

$$\boxed{\Delta E = \mu_0 B \left[1 + \frac{J(J+1) + S(S+1) - L(L+1)}{2J(J+1)}\right]M_J} \tag{3.15}$$

---

[3]The defining equation differs by a factor of c between SI and Gaussian units. The most convenient value is obtained by taking the Gaussian value and converting ergs to cm^{-1} to obtain $4.6685 \times 10^{-5}\ \frac{\text{cm}^{-1}}{\text{gauss}}$.

or

$$\Delta E = \mu_0 B g M_J \tag{3.16}$$

where g is the Landé g-value defined as

$$g \equiv 1 + \frac{J(J+1) + S(S+1) - L(L+1)}{2J(J+1)}$$

So a magnetic field removes the degeneracy associated with the isotropy of space. Note that the g-value is a property of the *level*.

Spectroscopy is a direct measurement of the energy differences between two different levels. Consider

$$E_1 = E_1^0 + \Delta E_1$$

$$\text{and} \quad E_2 = E_2^0 + \Delta E_2$$

Then the transition energy is just

$$T \equiv E_1 - E_2 = (E_1^0 - E_2^0) + \mu_0 B(g_1 M_{J_1} - g_2 M_{J_2})$$

There is a selection rule for transitions that demands that $M_{J_1} - M_{J_2} = 0, \pm 1$ but not $0 \to 0$.

Consider a transition $2s2p\,^3P_1 - 2p^2\,^3P_2$. In the absence of a magnetic field, this would produce a single spectral line with a transition energy of $T = E_1^0 - E_2^0$ and a wavelength of $\lambda = hc/T$ or $\lambda = 1/T$ if T is expressed in cm^{-1}. A magnetic field will split the line into how many components?

$$T = (E_1^0 - E_2^0) + \mu_0 B \frac{3}{2}(M_{J_1} - M_{J_2})$$

which follows since $g_1 = g_2 = {}^3/_2$.

Even though M_{J_1} can take on three values and M_{J_2} can take on five values, the previously mentioned selection rule demands that only three different transition energies (or spectral lines) are allowed. These are

$$T_1 = (E_1^0 - E_2^0) + \frac{3}{2}\mu_0 B$$

$$T_2 = (E_1^0 - E_2^0)$$

$$T_3 = (E_1^0 - E_2^0) - \frac{3}{2}\mu_0 B$$

Drawing the energy-level diagram will help you to see this.

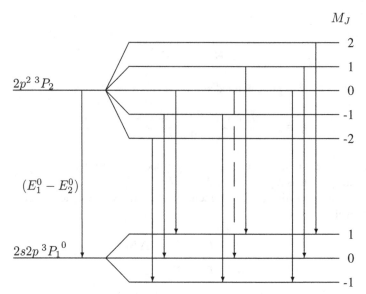

Splitting of a single spectral line into three components in the presence of an external magnetic field is called the normal Zeeman effect. Any time the g-values of the upper and lower states are the same, this pattern results. When the g-values are different, much more complicated patterns are possible. These, for historical reasons, are called the anomalous Zeeman effect.

Realize that the diagram above *is not to scale!* For an optical transition (5,000 Å), $E_1^0 - E_2^0 = 20{,}000$ cm^{-1}, while $^3\!/_2\mu_0 B = 0.7$ cm^{-1} for a magnetic field of 10,000 gauss (a typical value).

Problem 3.6

Find the Zeeman structure of a spectral line which results from the transition $^4F_{3/2} - ^4D_{5/2}$.

Return now to (3.14), which is

$$W = \mu_0(\mathbf{J}_z + \mathbf{S}_z)B$$

The matrix elements of $\mathbf{J}_z$ in the basis $LSJM_J$ are easy and are given by M_J for diagonal elements and zero otherwise. The matrix elements for $\mathbf{S}_z$ are not difficult to obtain for a given set of quantum numbers, but a general expression requires techniques not taken up in this text. The results are:

Diagonal elements of $\mathbf{S}_z$:

$$\langle L\,S\,J\,M_J|\mathbf{S}_z|L\,S\,J\,M_J\rangle = M_J\frac{[S(S+1)+J(J+1)-L(L+1)]}{2J(J+1)} \quad (3.17)$$

Off-diagonal elements:

$$\langle L\,S\,J\,M_J\,|\mathbf{S}_z|\,L\,S\,(J+1)\,M_J\rangle =$$
$$-\left[\frac{(J+M_J+1)(J-M_J+1)(S+L+J+2)}{(2J+3)(2J+2)^2}\times\right.$$
$$\left.\frac{(S+L-J)(S-L+J+1)(-S+L+J+1)}{(2J+1)}\right]^{1/2} \quad (3.18)$$

All other off-diagonal elements are zero. Note that use of (3.17), which is exact, in first-order perturbation theory, yields the Landé result. This is the quantum mechanical justification for that result.

Note further that the off-diagonal elements differ only by one in the J value. For example, if a $^3P_2$ differs in energy from $^3P_1$ (all other QN's the same) by much more than the magnetic interaction energy W, one can work in the subspace of a given L, S, and J. In that subspace $\mathbf{S}_z$ is diagonal and the Landé formula is applicable.

3.3 Stark Effect

An atom placed in an external electric field is subject to spectral line splitting called the Stark effect. This is similar to that experienced in an external magnetic field, but now the interaction is between the electric multipole moments of an atom with an applied, external electric field. The lowest such moment is the dipole moment, but an atom doesn't have a permanent dipole moment, so one expects that interaction to be zero. Using first-order perturbation theory, for complex atoms, that expectation is correct. Interestingly, because an atom has an electron cloud and that cloud need not be spherical, all multipole moments should be considered as possibly interacting with an external field, and one should let the wave functions decide whether or not the interaction is zero. Recall from your study of electrostatics that the interaction energy is the scalar product of the dipole moment with the derivative of the potential (the electric field). The interaction energy for higher multipole moments is the scalar product of the moment with higher derivatives of the potential. If the electric field is an applied laboratory field, these higher derivatives will be insignificant over the size of an atom, but if the field is caused by neighboring ions, that may not be the case. Nonetheless, only interactions with the dipole moment of the atom will be considered.

The dipole moment for an electron is just $-e\vec{r}$ and its expectation value for any atomic state is zero. That is simply a result of the fact that it is an odd parity operator and all states of an atom have definite parity. Hence, using first-order perturbation theory, the interaction with an external field is zero. The hydrogen atom presents an interesting special case. It has degenerate states of opposite parity for principal quantum numbers 2 or higher, so one can expect that this dipolar interaction will play a dominant role for hydrogen. This interaction will scale linearly with the applied electric field and is referred to as the *linear Stark effect* and is important for hydrogen even at low values of the applied field. The lowest-order interaction for complex atoms scales as the square of the applied electric field and is referred to as the *quadratic Stark effect*. Each of these will be developed in the following sections.

3.3.1 Linear Stark Effect

The Hamiltonian for a hydrogen atom in an external electric field, $\mathcal{E}$, may be written as

$$\mathbf{H} = \mathbf{H}_0 - e\vec{r} \cdot \vec{\mathcal{E}}$$

and letting $\vec{\mathcal{E}}$ lie along the z-axis yields

$$\mathbf{H} = \mathbf{H}_0 - ez\mathcal{E}.$$

This can be solved by direct diagonalization; consider the matrix

$$\langle n\ell m_\ell |\mathbf{H}| n\ell' m_{\ell'} \rangle.$$

Note that our attention here is restricted to a given n manifold and that spin is ignored. For a one-electron atom, all levels with a given n-value are degenerate, and this procedure of direct diagonalization is often taught as degenerate perturbation theory. One then obtains

$$\langle n\ell m_\ell |\mathbf{H}_0| n\ell' m_{\ell'} \rangle - e\mathcal{E}\langle n\ell m_\ell |r\cos\theta| n\ell' m_{\ell'} \rangle$$

$$= \epsilon_0 \delta_{\ell\ell'}\delta_{m_\ell m_{\ell'}} - e\mathcal{E}c^1(\ell m_\ell; \ell' m_{\ell'}) \int R_{n\ell}R_{n\ell'}r^3 \, dr$$

Here $c^1(\ell m_\ell; \ell' m_{\ell'})$ is the same c^k coefficient (k=1) as defined in Chap. 2. The only nonzero matrix element on the right for the $n = 2$ manifold is

$$\langle 200|r\cos\theta|210 \rangle = \langle 210|r\cos\theta|200 \rangle = -3r_B$$

Problem 3.7

Verify that $-3r_B$ is the value of the above matrix element.

The matrix to diagonalize is then

$$\begin{array}{ccccc}
2p & 2p & 2s & 2p & n\,\ell \\
1 & 0 & 0 & -1 & m_\ell
\end{array}$$

$$\begin{bmatrix}
\epsilon_0 & 0 & 0 & 0 \\
0 & \epsilon_0 & -3er_B\mathcal{E} & 0 \\
0 & -3er_B\mathcal{E} & \epsilon_0 & 0 \\
0 & 0 & 0 & \epsilon_0
\end{bmatrix}.$$

Clearly the $2p\ m_\ell = \pm 1$ are not changed in energy and still maintain the value ϵ_0. The determinantal equation for the 2×2 matrix in the center is

$$(\epsilon_0 - \lambda)^2 - 9e^2\mathcal{E}^2 r_B^2 = 0$$

$$\text{or} \quad \lambda = \epsilon_0 \pm 3er_B\mathcal{E}$$

Problem 3.8

By what amount will the $n = 2$ state of hydrogen be split if the atom is 20 a.u. from a proton?

Clearly the electric field mixes the $2s\ m_\ell = 0$ and $2p\ m_\ell = 0$ states and lifts (partially) the degeneracy among the $n = 2$ states. Inclusion of spin does not qualitatively change this result. For very strong fields, however, one should include interactions between differing n manifolds. This can be done by enlarging the size of the matrix, but note that then all of the diagonal energies are no longer the same. The off-diagonal element is much smaller than the energy difference among the diagonal elements and one obtains the quadratic Stark effect. The same result is obtained using second-order perturbation theory. For complex atoms the diagonal elements are not degenerate and the quadratic Stark effect is the lowest-order effect. The following problem illustrates these possibilities.

Problem 3.9

Consider a perturbation w (which could be an electric field interaction) for a two-level system with energies E_1 and E_2:

$$\begin{bmatrix} E_1 & w \\ w & E_2 \end{bmatrix}$$

(a) Obtain the exact solution for the eigenvalues of this matrix.
(b) Find an approximation valid whenever $w \ll (E_1 - E_2)$.

3.3.2 Quadratic Stark Effect

There are two ways to obtain a result that scales like the square of the applied external field. The first is to consider all states whose energies are close to the level of interest, evaluate the matrix for those states, and then diagonalize it. The problem above exemplifies that method for a two-level system. The second is to employ second-order perturbation theory, which in principle involves an infinite sum but in practice the sum is truncated so as to include only those levels close in energy to the one of interest. By "close" I mean comparable to the magnitude of the perturbation, but even that varies a lot depending on the problem at hand and on the precision desired for the result.

The Hamiltonian, for a multielectron atom in an external electric field, is similar to what was written previously:

$$\mathbf{H} = \mathbf{H}_0 - \sum_i e\vec{r}_i \cdot \vec{\mathcal{E}}$$

The operator is best converted to a spherical tensor. Again letting the electric field lie along the z-axis produces

$$-\sum_i e\vec{r}_i \cdot \vec{\mathcal{E}} = -\sum_i \mathcal{E}\, e\, r_i Y_1^0(\theta_i, \phi_i) \equiv \mathcal{E} \sum_i D_1^0(r_i, \theta_i, \phi_i) \qquad (3.19)$$

D_1^m is the spherical tensor form of the dipole moment operator. What is desired here is the matrix element of a one-electron operator which, from Chap. 2, can be written without reference to any electron indices. The sum over electrons becomes a sum over quantum numbers. Specifically, using second-order perturbation theory, it is straightforward to write the perturbed energy as

$$E'_{\gamma,J,M} = \mathcal{E}^2 \sum_{\gamma',J',M'} \frac{\langle \gamma J M | D_1^0 | \gamma' J' M' \rangle \langle \gamma' J' M' | D_1^0 | \gamma J M \rangle}{E_{\gamma,J,M} - E_{\gamma',J',M'}} \qquad (3.20)$$

Here E' is the perturbed energy, while E is the unperturbed energy; γ represents all of the unwritten quantum numbers, specifically the configuration and total orbital and spin quantum numbers. Invoking the Wigner–Eckart theorem, this may be rewritten as

$$E'_{\gamma,J,M} = \mathcal{E}^2 \sum_{\gamma',J'} (-1)^{J+J'} \begin{pmatrix} J' & J & 1 \\ M & -M & 0 \end{pmatrix}^2 \frac{|\langle \gamma J \| D_1 \| \gamma' J' \rangle|^2}{E_{\gamma,J} - E_{\gamma',J'}} \qquad (3.21)$$

Problem 3.10

Obtain Eq. (3.21) from (3.20).

This expression, while quite general, is not so easy to apply as others seen previously. The reduced matrix element would need to be evaluated in the basis of one-electron spin orbitals, and learning to do that is subject matter for a more advanced course. Nonetheless, a few conclusions can be drawn. The matrix elements are zero unless the states have opposite parity. The ground state for most atoms is well separated in energy from the first excited state, so one can conclude that the quadratic Stark effect would only be significant for excited states of atoms or ions subjected to external, laboratory-scale, electric fields.

3.4 Hyperfine Structure

Hyperfine structure, so named because spectral lines show a splitting or structure at very high resolution whenever the nuclear spin is nonzero, is a consequence of the interactions of the atomic electron cloud with the magnetic and electric multipole moments of the nucleus.[4] No interactions occur for closed shells or subshells. The leading-order interaction occurs between the total atomic angular momentum, J, and the magnetic dipole moment of the nucleus given by

$$\vec{\mu_I} = \mu_n \, g_I \, \vec{I}. \tag{3.22}$$

This expression is completely analogous to the one written previously for the electron, Eq. (3.12). Here $\vec{\mu_I}$ is the nuclear magnetic moment. $\vec{I}$ is the total angular momentum of the nucleus. μ_n is the nuclear magneton which is the Bohr magneton with the electron mass replaced with the proton mass, i.e.,

$$\mu_n = \mu_0 \frac{m_e}{m_p}$$

What is significantly different from the case for electrons is that the nuclear g-factor, g_I, ranges from $-4.3(^3\text{He})$ to $+5.3(^{19}\text{F})$, while nuclear spins range from 0 to 7 (^{176}Lu).[5] The nuclear spin, $\vec{I}$, couples to the total electronic angular momentum, $\vec{J}$, to form $\vec{F}$ the coupled angular momentum which satisfies all of our rules for angular momentum coupling. Explicitly,

$$\vec{F} = \vec{J} + \vec{I}. \tag{3.23}$$

The interaction energy, just to the level of the nuclear magnetic dipole, is not so easy to write down. If one thinks of a magnetic field caused by the electron cloud pointing in the direction of $\vec{J}$, one might write

---

[4]See L. Armstrong, Jr. *Theory of the Hyperfine Structure of free Atoms* Wiley-Interscience, New York, 1971.

[5]Robert D. Cowan, *The Theory of Atomic Structure and Spectra*, Univ. of Calif. Press, Berkeley, 1981.

$$\vec{B}_J = \frac{\vec{B} \cdot \vec{J}}{J^2} \vec{J}. \tag{3.24}$$

Then the magnetic hyperfine structure Hamiltonian could be written as

$$H_{mhfs} = -\vec{B}_J \cdot \vec{\mu}_I = -\mu_n g_I \frac{\vec{B} \cdot \vec{J}}{J^2} \vec{J} \cdot \vec{I} \equiv A \vec{J} \cdot \vec{I} \tag{3.25}$$

The matrix element that yields the perturbed energy in first-order perturbation theory is just

$$E' = \langle \gamma JIF | A \vec{J} \cdot \vec{I} | \gamma JIF \rangle = \tfrac{1}{2} A[F(F+1) - J(J+1) - I(I+1)] = \tfrac{1}{2} AK \tag{3.26}$$

where

$$K \equiv F(F+1) - J(J+1) - I(I+1). \tag{3.27}$$

Problem 3.11

Show that two energy levels differing only by one in their value for F have a separation given by AF where F is the larger of the two.

A represents the actual interaction energy and is not simple to calculate. It is often inferred from measurements. For the case of one electron outside of a closed shell, the A value can be calculated from expressions that differ depending on whether the electron is or is not in an s-state,[6] For $\ell > 0$,

$$A = \mu_0 \mu_n \frac{2\ell(\ell+1)}{J(J+1)} g_I < r^{-3} >_{n\ell} . \tag{3.28}$$

For $\ell = 0$,

$$A = \frac{16\pi}{3} \mu_0 \mu_n g_I |\psi(0)|^2. \tag{3.29}$$

Here $\psi(0)$ is the wave function for the single outer electron at the origin, the angular momentum quantum numbers are dimensionless, and the expressions are using Gaussian units. Equation (3.29) was first worked out by Enrico Fermi and this term is called the Fermi contact term. It's useful to realize that when there is an unpaired s electron, this interaction is the largest of the hyperfine terms.

---

[6]E. Arimondo, M. Inguscio, and P. Violino, Rev.Mod.Phys. **49, 31 (1977)**.

If one now includes the nuclear electric quadrupolar interaction with the electron cloud, the hyperfine interaction can be written as

$$E'_{hfs} = \tfrac{1}{2}AK + B\frac{[\tfrac{3}{2}K(K+1) - 2I(I+1)J(J+1)]}{2I(2I-1)2J(2J-1)}. \qquad (3.30)$$

B is given by

$$B = e^2\frac{(2J-1)}{(2J+2)} <r^{-3}>_{n\ell} Q \qquad (3.31)$$

where Q is the nuclear quadrupole moment.

The situation in which there is more than one electron outside of a closed shell is significantly more complicated. There are also higher-order interactions which cannot necessarily be ignored. Again, helium (^{3}He) is a good example of a common element in which the above expressions are insufficient for reasonable accuracy. Theoretical expressions have been worked out[7] and an example of their application is also available.[8] Figure 3.1 shows the energy-level positions for the 1s3d (all of the 1snd are similar) of atomic ^{3}He. The singlet–triplet separation, called electrostatic exchange, was treated in the previous chapter. Note the relative size of the fine structure and hyperfine structure for this manifold. That the hyperfine structure is comparable in size to the fine structure occurs only for helium. Its fine structure is particularly small and the Fermi contact term, $-8,667.8$ MHz (^{3}He$^+$), caused by an unpaired 1s electron, is significantly larger than the fine structure. The contact term appears in both diagonal and off-diagonal matrix elements, so no elementary treatment comes close to approximating these energy separations.

Perhaps the most widely known and arguably most important example of hyperfine structure occurs in the ground state of atomic hydrogen. The nucleus is a proton with $I = \tfrac{1}{2}$ and the ground state, $1s_{1/2}$, has $J = \tfrac{1}{2}$. Hence F may equal 0 or 1. The transition between those two levels is referred to as the 21 cm line of hydrogen by astronomers which is nothing but the energy difference expressed in wavelength. Can this calculation yield a value close to that for this transition?

The g_I factor for the proton is $+5.5856$. Invoking the result of a recent problem, the answer to the question of what is the energy separation between the lowest hyperfine levels of atomic hydrogen will be given by evaluating Eq. (3.29). While this is not difficult, it becomes an exercise in understanding Gaussian units.

Problem 3.12

Show that μ_0^2/a_0^3 is given by $\tfrac{1}{4}\alpha^2$ Hartree where μ_0 is the Bohr magneton, a_0 is the Bohr radius, and α is the fine structure constant.

---

[7] A. Lurio, M. Mandel and R. Novick, Phys.Rev. **126, 1758 (1962)**.
[8] R.L. Brooks, V.F. Streif and H.G. Berry, Nucl. Instrum. Meth. **202, 113 (1982)**.

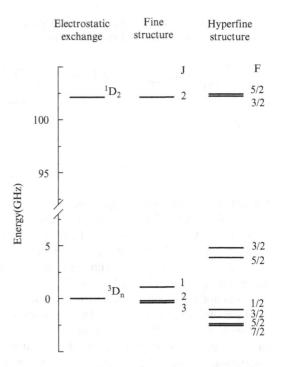

Fig. 3.1 Energy-level diagram of 1s3d ^{3}He I

If one looks up the ground-state radial wave function for hydrogen from Chap. 1 and evaluates it at $r = 0$, one obtains a value of 2. Two what? Remember that the square of the wave function integrated over all space must equal 1, unitless, but dV in the integrand has dimensions of distance cubed. Hence, the square of the wave function itself must have both the dimensions and units of a_0^{-3} and the wave function itself of $a_0^{-3/2}$. Next you have to remember that the angular part of the wave function is given by Y_0^0. With all of these comments and the result of the problem given above, you should be able to show that the energy separation between the two hyperfine components of the ground state of hydrogen is 0.04738 cm^{-1}, yielding a wavelength of 21.1 cm and a frequency of 1,420 MHz.

Problem 3.13

Obtain the result stated in the previous paragraph.

Problem 3.14

Perform the same calculation for ^{3}He$^+$, using a value for g_I of -4.255, and see how close to $-8,668$ MHz you can come.

Chapter 4
Transition Probabilities

In order to develop the concept of transitions between excited states of atoms based completely on the formalism of quantum mechanics, time-dependent perturbation theory will be described and then applied to the problem of spontaneous emission of an excited atomic state. Every author faces a dilemma when trying to do this at an undergraduate level. The difficulty is that spontaneous emission can most properly be thought of as an interaction of the quantized radiation field with an atom. To do that interaction properly one should quantize the radiation field and apply second quantization to the atom. All of that is beyond the scope of this text, so what is to be done? Many authors choose to do a semiclassical treatment in which a classical radiation field interacts quantum mechanically with an atom. The level of detail varies considerably. This presentation will attempt to be complete, explicitly stating where a result from more advanced quantum mechanics is invoked, even as that result is explained and motivated.

4.1 Time-Dependent Perturbation Theory

The time-dependent Schrödinger equation describes the time evolution of the state of a physical system and is given by

$$\mathbf{H}\psi(t) = i\hbar \frac{\mathrm{d}}{\mathrm{d}t}\psi(t) \tag{4.1}$$

At some time t_0 (and for all earlier times), assume that the state of the system satisfies the time-independent equation

$$\mathbf{H}\psi(t_0) = E\psi(t_0) \tag{4.2}$$

If $\mathbf{H}$ is time independent, (4.1) can be integrated to yield

R.L. Brooks, *The Fundamentals of Atomic and Molecular Physics*, Undergraduate Lecture Notes in Physics, DOI 10.1007/978-1-4614-6678-9_4, © Springer Science+Business Media New York 2013

$$\psi(t) = e^{-i\mathbf{H}(t-t_0)/\hbar}\psi(t_0) \tag{4.3}$$

This solution can be readily verified by differentiation.

$$\frac{d}{dt}\psi(t) = \frac{-i\mathbf{H}}{\hbar}e^{-i\mathbf{H}(t-t_0)/\hbar}\psi(t_0)$$

$$= \frac{-i\mathbf{H}}{\hbar}\psi(t)$$

$$\text{or} \quad i\hbar\frac{d}{dt}\psi t = \mathbf{H}\psi(t) \tag{4.1}$$

What does it mean to have an operator in the exponent as in (4.3)? It means just a series expansion as

$$e^{\mathbf{A}t} = 1 + \mathbf{A}t + \frac{\mathbf{A}^2 t^2}{2} + \cdots$$

So when $\mathbf{H}$ is time independent, (4.3) gives

$$\psi(t) = e^{-iE(t-t_0)/\hbar}\psi(t_0) \tag{4.4a}$$

$$\text{or} \quad \psi(t) = e^{-iEt/\hbar}\psi(0) \tag{4.4b}$$

Of course, since (4.2) is an eigenvalue equation the above is valid for each eigenvector k, or

$$\psi_k(t) = e^{-iE_k t/\hbar}\psi_k(0) \tag{4.5}$$

If the system were prepared in the kth eigenstate at time $t = 0$, the probability of finding it in the mth state at a later time is

$$|\langle \psi_m(t) \mid \psi_k(0) \rangle|^2 = \left| e^{-iE_m t/\hbar}\langle \psi_m(0) \mid \psi_k(0) \rangle \right|^2 = \delta_{mk}$$

So when a Hamiltonian is time independent, a system prepared in a given state stays in that state. This is the justification for treating the time-independent problem by a separate formalism. Of course, an atom in an excited state decays spontaneously which does not occur for the Hamiltonian previously considered. Hence, the Hamiltonian previously considered must be incomplete. There must be some time dependence that has yet to be considered.

Consider next the problem given by the equation

$$(\mathbf{H}_0 + \mathbf{H}_I)\psi(t) = i\hbar\frac{d}{dt}\psi(t) \tag{4.6}$$

$\mathbf{H}_I$ is a perturbation term in the Hamiltonian which will be time dependent, but only the time-independent solution is known and is given by

$$\mathbf{H}_0 u_k(\vec{r}) = E_k u_k(\vec{r}) \tag{4.7}$$

So u_k are the time-independent wave functions for the unperturbed Hamiltonian. Since these represent a complete set, $\psi(t)$ can be expanded as

$$\psi(t) = \sum_k C_k(t) u_k(\vec{r}) e^{-iE_k t/\hbar} \qquad (4.8)$$

The exponential factor represents the only time dependence $\mathbf{H}_0$ can produce.
  Putting (4.8) into (4.6) gives

$$(\mathbf{H}_0 + \mathbf{H}_I)\psi(t) = i\hbar \sum_k \left[\dot{C}_k(t) u_k e^{-iE_k t/\hbar} - \frac{iE_k}{\hbar} C_k(t) u_k e^{-iE_k t/\hbar} \right]$$

The second term on the RHS is just $\mathbf{H}_0\psi(t)$, leaving

$$\sum_k \mathbf{H}_I C_k u_k e^{-iE_k t/\hbar} = i\hbar \sum_\ell \dot{C}_\ell(t) u_\ell e^{-iE_\ell t/\hbar}. \qquad (4.9)$$

Because the two sums are independent, the index on the right has been changed to ℓ. Now multiply both sides by $u_m^* e^{iE_m t/\hbar}$, integrate over all space, and interchange the left- and right-hand sides:

$$i\hbar \sum_\ell \dot{C}_\ell(t) \delta_{\ell m} e^{i(E_m - E_\ell)t/\hbar} = \sum_k \langle m|\mathbf{H}_I|k\rangle e^{i(E_m - E_k)t/\hbar} C_k(t) \qquad (4.10)$$

The sum over ℓ picks out the mth component of $\dot{C}_\ell$ giving

$$\dot{C}_m(t) = \frac{1}{i\hbar} \sum_k \langle m|\mathbf{H}_I|k\rangle e^{i(E_m - E_k)t/\hbar} C_k(t) \qquad (4.11)$$

where

$$\langle m|\mathbf{H}_I|k\rangle \equiv \int u_m^*(\vec{r}) \mathbf{H}_I u_k(\vec{r}) \, d\tau.$$

  In Eq. (4.11) which is exact, k is a dummy index that is summed over. Here m is the index of interest. One can solve this equation iteratively by letting

$$C_m(t) = C_m^{(0)}(t) + C_m^{(1)}(t) + C_m^{(2)}(t) + \cdots \qquad (4.12)$$

Starting with the assumption that the zero-order C coefficient is unity for *some* index but zero for all other indices, letting it be unity when the index is ℓ yields $C_\ell^0(t) = 1$ and $C_k^0(t) = 0$ for $k \neq \ell$; then Eq. (4.11) can be rewritten as

$$\dot{C}_m^{(1)}(t) = \frac{1}{i\hbar} \sum_k \langle m|\mathbf{H}_I|k\rangle e^{i(E_m - E_k)t/\hbar} C_k^{(0)}(t). \qquad (4.13)$$

Integrating over time and remembering that only one term in the sum is nonzero gives the first order result:

$$C_m^{(1)}(t) = \frac{1}{i\hbar} \int_0^t dt' \langle m | \mathbf{H}_I | \ell \rangle e^{i(E_m - E_\ell)t'/\hbar}. \tag{4.14}$$

Then to this order $C_m(t) = C_m^{(1)}(t)$ (of course $C_\ell(t) = 1$). What has been done here is to prepare the system at time $t = 0$ in the ℓth state and to let it evolve in time.

The second-order term is

$$C_m^{(2)}(t) = \frac{1}{i\hbar} \sum_n \int_0^t dt'' \langle m | \mathbf{H}_I(t'') | n \rangle e^{i(E_m - E_n)t''/\hbar} C_n^{(1)}(t'')$$

$$= \frac{1}{i\hbar} \sum_n \int_0^t dt'' \int_0^{t''} dt' \langle m | \mathbf{H}_I(t'') | n \rangle e^{i(E_m - E_n)t''/\hbar}$$

$$\langle n | \mathbf{H}_I(t') | \ell \rangle e^{i(E_n - E_\ell)t'/\hbar} \tag{4.15}$$

and so on.

Note that this second-order result has an intermediate, infinite sum not present in the first-order result. For that reason, second-order, time-dependent perturbation theory is not so commonly used. With the formalism completed, let us consider how to apply perturbation theory to atomic transitions. Our starting point is to derive an expression valid to first order for any harmonic time-dependent perturbation. If for any reason $\mathbf{H}_I(t) = \mathbf{H}_I' e^{\mp i\omega t}$ with $\mathbf{H}_I'$ independent of time,

$$C_m^{(1)}(t) = \frac{1}{i\hbar} \langle m | \mathbf{H}_I' | \ell \rangle \int_0^t dt' e^{i(E_m - E_\ell \mp \hbar\omega)t'/\hbar}. \tag{4.16}$$

The $C_k(t)$ are the coefficients in an orthonormal expansion of $\psi(t)$. Since $\langle \psi(t) | \psi(t) \rangle = 1$, the sum of the squares of the coefficients is unity. So $|C_m|^2$ is interpreted as the probability of finding the system in the mth state.

Return now to (4.16). Let

$$(E_m - E_\ell \mp \hbar\omega)/\hbar \equiv \omega_{m\ell} \mp \omega \equiv \Omega. \tag{4.17}$$

The integral over t' can be performed and yields

$$\frac{1}{i\Omega} e^{i\Omega t'} \Big]_0^t = \frac{1}{i\Omega} [e^{i\Omega t} - 1] = \frac{2e^{i\Omega t/2}}{\Omega} \left[\frac{e^{i\Omega t/2} - e^{-i\Omega t/2}}{2i} \right] = \frac{2e^{i\Omega t/2}}{\Omega} \sin(\Omega t/2).$$

Then

$$C_m^{(1)}(t) = \frac{2e^{i\Omega t/2}}{i\hbar\Omega} \sin(\Omega t/2)\langle m\,|\mathbf{H}_I'|\,\ell\rangle \tag{4.18}$$

$$\left|C_m^{(1)}(t)\right|^2 = \frac{4}{\hbar^2}\frac{\sin^2(\Omega t/2)}{\Omega^2}|\langle m\,|\mathbf{H}_I'|\,\ell\rangle|^2. \tag{4.19}$$

It is tempting to try to interpret this equation as an oscillatory change in probability between the original state ℓ and the state of interest m. Something like that occurs when the light field is considered as the source of the interaction but the intensity is so high that a perturbative treatment, like this one, is insufficient to explain the observed phenomena. In our treatment it is more fruitful to think of the perturbation as weak, in which case only resonant interactions (ones for which $\Omega = 0$) are of interest.

Often this is as far as such an expression is carried. However, note that

$$\frac{4\sin^2(\Omega t/2)}{\hbar^2\Omega^2} = \frac{2\pi t}{\hbar^2}\left[\frac{\sin^2(\Omega t/2)}{\pi(t/2)\Omega^2}\right].$$

If the transition probability per unit time is defined as

$$P_{\ell\to m} \equiv \lim_{t\to\infty} \frac{\left|C_m^{(1)}(t)\right|^2}{t}$$

$$= \frac{2\pi}{\hbar^2}|\langle m\,|\mathbf{H}_I'|\,\ell\rangle|^2 \lim_{t\to\infty}\left[\frac{\sin^2(\Omega t/2)}{\pi(t/2)\Omega^2}\right]$$

then the limit can be recognized as one definition of the Dirac delta function:

$$\lim_{\alpha\to\infty}\frac{1}{\pi}\frac{\sin^2\alpha x}{\alpha x^2} = \delta(x). \tag{4.20}$$

In this way the transition probability in going from the ℓth state to the mth is given by an expression often referred to as Fermi's golden rule:

$$\boxed{P_{\ell\to m} = \frac{2\pi}{\hbar^2}|\langle m\,|\mathbf{H}_I'|\,\ell\rangle|^2\,\delta(\omega_{m\ell}\mp\omega).} \tag{4.21}$$

The Dirac delta function, more properly a distribution, has meaning only inside of an integral expression. It is equal to zero everywhere except at zero where it is undefined (i.e., infinite). Its role is to eliminate all contributions of the integrand except those at $x = 0$ (or at the zero of the argument of the delta function). In other words,

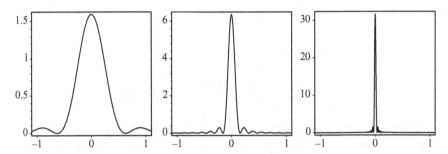

Fig. 4.1 Argument of the limit in (4.20) for $\alpha = 5, 20$ and 100

$$f(x_0) = \int_{x<x_0}^{x>x_0} f(x)\, \delta(x - x_0)\, \mathrm{d}x.$$

The Dirac delta function is normalized to unity. That is,

$$1 = \int_{x<x_0}^{x>x_0} \delta(x - x_0)\, \mathrm{d}x.$$

One can graphically see how the expression in Eq. (4.20) becomes a spike with vanishingly narrow x-extent by looking at that expression for three different values of α (Fig. 4.1).

4.2 Spontaneous Emission

The previous development assumed a harmonic interaction Hamiltonian. To see how such a Hamiltonian might arise consider the classical Hamiltonian for the ith particle in an electromagnetic field given by

$$\mathbf{H} = \frac{1}{2m}\left(\vec{p}_i - \frac{q}{c}\vec{A}_i\right)^2 + q\Phi_i \qquad (4.22)$$

where $\vec{A}_i$ and Φ_i are the vector and scalar potentials of the field at the position of the ith particle. Further, recall that any vector $\vec{A}$ can be decomposed into transverse and longitudinal components

$$\vec{A} = \vec{A}_\perp + \vec{A}_\|$$

such that $\vec{\nabla} \cdot \vec{A}_\perp = 0$ and $\vec{\nabla} \times \vec{A}_\| = 0$. The vector potential in the Hamiltonian being considered will be treated as a transverse vector allowing for a somewhat simpler treatment of what follows. This is equivalent to working in the Coulomb gauge which is appropriate for a large class of light–matter interactions.

The potential Φ in the above Hamiltonian may be interpreted as any electrostatic potential or indeed as any superposition of electrostatic potentials from any number of sources. For a particle in the field of the nucleus, $\Phi_i = e/r_i$. If there was also an external constant electric field, then from the superposition principle, $\Phi_i = e/r_i - E_0 z$. For a system of many charged particles (electrons) in the electrostatic field of the nucleus and each other, the superposition principle simply demands one term being the electrostatic attraction to the nucleus and a second term involving the electrostatic repulsion of each electron to any other. Then the Hamiltonian of (4.22) can be written as

$$\mathbf{H} = \sum_{i=1}^{N} \left[\frac{1}{2m} \left(\vec{p}_i - \frac{e}{c} \vec{A}_i \right)^2 - \frac{Ze^2}{r_i} \right] + \sum_{i>j}^{N} \frac{e^2}{r_{ij}} \tag{4.23}$$

where $\vec{A}_i$ should be interpreted as $\vec{A}_\perp(\vec{r}_i, t)$. This guarantees that $\operatorname{div} \vec{A} = 0$, a criterion for working in the Coulomb gauge. This Hamiltonian can be expanded into two parts, one which was treated previously and another called the interaction Hamiltonian:

$$\mathbf{H} = \mathbf{H}_0 + \mathbf{H}_I$$

with

$$\mathbf{H}_0 = \sum_{i=1}^{N} \left(\frac{1}{2m} \vec{p}_i^2 - \frac{Ze^2}{r_i} \right) + \sum_{i>j}^{N} \frac{e^2}{r_{ij}} \tag{4.24}$$

$$\mathbf{H}_I = \sum_{i=1}^{N} \left\{ -\frac{e}{2mc} \left[\vec{p}_i \cdot \vec{A}(\vec{r}_i, t) + \vec{A}_i \cdot \vec{p}_i \right] + \frac{e^2}{2mc^2} \vec{A} \cdot \vec{A} \right\} \tag{4.25}$$

Recall that $\vec{p} = i\hbar\vec{\nabla}$ and the gradient operator operates on everything to its right, so for any ϕ

$$\vec{\nabla} \cdot \vec{A}\phi = \phi\vec{\nabla} \cdot \vec{A} + \vec{A} \cdot \vec{\nabla}\phi$$

But $\vec{\nabla} \cdot \vec{A} = 0$ so $\vec{p} \cdot \vec{A} = \vec{A} \cdot \vec{p}$.

Because the vector potential represents the perturbation, which in this treatment is assumed to be small, the term $\vec{A} \cdot \vec{A}$ will be dropped and the interaction Hamiltonian becomes

$$\mathbf{H}_I = \sum_{i=1}^{N} -\frac{e}{mc} \left(\vec{A}(\vec{r}_i, t) \cdot \vec{p}_i \right) \tag{4.26}$$

Consider the problem of an atom in a bath of electromagnetic radiation composed of near optical frequencies (say far infrared to ultraviolet). Classically one considers the source of this bath of radiation to be external to the atom so that

$$\nabla^2 \vec{A} - \frac{1}{c^2}\frac{\partial^2 \vec{A}}{\partial t^2} = 0$$

$$\vec{E} = \frac{1}{c}\frac{\partial \vec{A}}{\partial t}$$

$$\vec{B} = \vec{\nabla} \times \vec{A}$$

Then the plane wave solution for $\vec{A}$ must be of the form

$$\vec{A}(\vec{r}, t) = \vec{A}_0 e^{\pm i(\vec{k}\cdot\vec{r} - \omega t)} \tag{4.27}$$

with

$$\vec{A}_0 \cdot \vec{k} = 0 \tag{4.28}$$

Before an expression like (4.27) can be put into (4.26), one needs to know how to convert $\vec{A}$ into a quantum mechanical operator. This entails quantizing the radiation field and the derivation would take us too far into quantum electrodynamics, but the result is easy to understand.

For absorption,

$$\vec{A} \rightarrow c\sqrt{\frac{2\pi\hbar n_k}{\omega V}}\,\vec{\epsilon}\,e^{i(\vec{k}\cdot\vec{r} - \omega t)} \tag{4.29}$$

For emission,

$$\vec{A} \rightarrow c\sqrt{\frac{2\pi\hbar(n_k + 1)}{\omega V}}\,\vec{\epsilon}\,e^{-i(\vec{k}\cdot\vec{r} - \omega t)} \tag{4.30}$$

Here n_k is the number of photons having wavevector $\vec{k}$ in a volume V, and $\vec{\epsilon}$ is a unit (polarization) vector at right angles to $\vec{k}$. ω is the frequency of the photon at position $\vec{r}$ traveling in the direction of the wavevector $\vec{k}$. Recall that $\vec{B} \cdot \vec{B}$ is proportional to an energy density. Since $\vec{B} = \text{curl}\,\vec{A}$, A^2 must have the dimensions of energy density times distance squared. Those are the dimensions of the above expression. One can think of $\vec{A}$ as representing a bath of electromagnetic radiation interacting with the atom of interest. Its "intensity" is represented by n_k/V. What is fascinating is that if the electromagnetic radiation is turned off, $\vec{A}$ is not zero but n_k is zero. Absorption is impossible but emission is still possible. This is a purely quantum electrodynamic effect resulting from the zero-point energy of the quantized electromagnetic field. How does one interpret $\vec{k}, \omega$, etc., when there are no photons? Since this only occurs for emission, one interprets these as the parameters of the emitted photon. How this comes about will be shown directly.

Let us consider the problem of spontaneous emission. One can then use (4.30) with $n_k = 0$. Putting that into (4.26) gives

$$\mathbf{H}_I = \sum_{i=1}^{N} -\frac{e}{m}\sqrt{\frac{2\pi\hbar}{\omega V}}(\vec{\epsilon}\cdot\vec{p}_i)e^{-i\vec{k}\cdot\vec{r}_i + i\omega t} \tag{4.31}$$

This will shortly be put into (4.21) of the previous section. There the expression was the probability per unit time that a transition would take place between two energy levels if the condition

$$\frac{E_m - E_\ell}{\hbar} + \omega = 0 \tag{4.32}$$

is satisfied, where E_ℓ is the energy of the initial state.

Equation (4.21) previously is

$$P_{\ell \to m} = \frac{2\pi}{\hbar^2} |\langle m | \mathbf{H}'_I | \ell \rangle|^2 \, \delta(\omega_{m\ell} \mp \omega) \tag{4.21}$$

This expression, however, needs to be multiplied by the number of states (photon states) having a frequency in the range ω to $\omega + d\omega$. Apply periodic boundary conditions to a box of arbitrary volume V such that

$$\frac{2\pi n}{L} = k_x \quad \text{or} \quad k_y \quad \text{or} \quad k_z \tag{4.33}$$

and $V = L^3$ and $n = \pm 1, \pm 2, \pm 3 \ldots \infty$. The number of allowed states is n^3. The density of states is formed by considering

$$\Delta n(\vec{k}) = \frac{L_x L_y L_z}{(2\pi)^3} \Delta k_x \Delta k_y \Delta k_z$$

$$\text{or} \quad dn(\vec{k}) = \frac{V \, d^3\vec{k}}{(2\pi)^3} \quad \text{(Density of states)}$$

$$|\vec{k}| = \frac{\omega}{c}.$$

For a spherical shell, $d^3\vec{k} \Rightarrow 4\pi k^2 \Delta k$; then

$$\Delta n(\vec{k}) = \frac{V 4\pi k^2 \Delta k}{(2\pi)^3}$$

Changing to frequency yields

$$\Delta n(\omega) = \frac{V 4\pi \omega^2 \Delta \omega}{(2\pi)^3 c^3} \Rightarrow \frac{V 4\pi \omega^2}{(2\pi)^3 c^3} \, d\omega$$

(Two states of polarization will be counted later.)

The result needs to be averaged over all angles which is accomplished by integrating over Ω and dividing by 4π. The integrand part of this can be inserted now by multiplying by $d\Omega/4\pi$.

The number of allowed states per unit frequency in a solid angle $d\Omega$ is then

$$\rho_{\omega,\,d\Omega}\,d\omega = \frac{V}{(2\pi)^3}\frac{\omega^2}{c^3}\,d\Omega\,d\omega \tag{4.34}$$

So the transition probability per unit time into a solid angle $d\Omega$, labeled $w_{d\Omega}$, must be

$$
\begin{aligned}
w_{d\Omega} &= \int P_{\ell\to m}\,\rho_{\omega,\,d\Omega}\,d\omega \\[4pt]
&= \frac{2\pi}{\hbar^2}|\langle\,m\,|\mathbf{H}'_I|\,\ell\,\rangle|^2\,\rho_{\omega,\,d\Omega} \\[4pt]
&= \frac{2\pi}{\hbar^2}|\langle\,m\,|\mathbf{H}'_I|\,\ell\,\rangle|^2\,\frac{V}{(2\pi)^3}\frac{\omega_{m\ell}^2}{c^3}\,d\Omega
\end{aligned}
\tag{4.35}
$$

Here one sees that the integral over ω has removed the delta function and turned the variable of integration, ω, into $\omega_{m\ell}$. This is the frequency determined by the energy difference between the upper and lower states. It is the frequency of the emitted photon. *In subsequent equations it will be written as ω rather than $\omega_{m\ell}$ since that is the only frequency left in the expressions.* Reflect on this point because it can be confusing to see a final expression having within it what you remember to be a variable of integration.

Writing out the matrix element, the transition probability may be written as

$$
\begin{aligned}
w_{d\Omega} &= \frac{2\pi}{\hbar^2}\frac{e^2}{m^2}\frac{2\pi\hbar}{\omega V}\left|\sum_i\langle\,m\,|e^{-i\vec{k}\cdot\vec{r}_i}\,\vec{\epsilon}\cdot\vec{p}_i|\,\ell\,\rangle\right|^2\frac{V\omega^2}{(2\pi)^3c^3}\,d\Omega \\[6pt]
&= \frac{e^2\omega}{2\pi\hbar m^2 c^3}\left|\sum_i\langle\,m\,|e^{-i\vec{k}\cdot\vec{r}_i}\,\vec{\epsilon}\cdot\vec{p}_i|\,\ell\,\rangle\right|^2 d\Omega.
\end{aligned}
\tag{4.36}
$$

Since $\lambda_{\text{photon}} = 2\pi/|\vec{k}| \gg r_{\text{atom}}$ as $r \approx 1$ Å and $\lambda \approx 5{,}000$ Å, the exponential above can be expanded as

$$e^{-i\vec{k}\cdot\vec{r}_i} = 1 - i\vec{k}\cdot\vec{r}_i - \frac{(\vec{k}\cdot\vec{r}_i)^2}{2} + \cdots \tag{4.37}$$

and replaced by the leading term of 1. This leaves us with an operator of the form $\sum_i \vec{\epsilon}\cdot\vec{p}_i$. This approximation is called the electric dipole approximation for reasons that will shortly be apparent. In fact, each term in the series is responsible for electric multipole radiation which will not concern us presently. This yields

$$w_{d\Omega} = \frac{e^2\omega}{2\pi\hbar m^2 c^3}\left|\sum_i\langle\,m\,|\vec{p}_i|\,\ell\,\rangle\cdot\vec{\epsilon}\right|^2 d\Omega. \tag{4.38}$$

Using the commutation relation

$$[\vec{p}_i^2, \vec{r}_i] = -2i\hbar\vec{p}_i$$

this may be written as

$$\sum_i \vec{p}_i = \sum_i \frac{i}{2\hbar}[\vec{p}_i^2, \vec{r}_i] = \frac{im}{\hbar}\left[\mathbf{H}_0, \sum_i \vec{r}_i\right].$$

This follows for two reasons. The first is that $\vec{r}_i$ commutes with $1/r_i$ and with $1/r_{ij}$ in $\mathbf{H}_0$. The second is that $\vec{r}_i$ commutes with $\vec{p}_j$ for $i \neq j$. Hence the operator becomes

$$\sum_i \langle m|\vec{p}_i|\ell\rangle = \frac{im}{\hbar}(E_m - E_\ell)\sum_i \langle m|\vec{r}_i|\ell\rangle$$

$$= im\omega\sum_i \langle m|\vec{r}_i|\ell\rangle \tag{4.39}$$

Then

$$w_{d\Omega} = \frac{\omega^3}{2\pi\hbar c^3}\left|\sum_i \langle m|e\vec{r}_i \cdot \vec{\epsilon}|\ell\rangle\right|^2 d\Omega \tag{4.40}$$

$e\vec{r}_i$ is classically a dipole moment, hence the name electric dipole radiation. All of the derivation following the expansion of the exponential in (4.37) carries through for any of the terms. Subsequent ones are simply multiplied by higher powers of $\vec{k}\cdot\vec{r}$. These give rise to electric quadrupole, electric octupole, etc., transitions which are many orders of magnitude weaker than electric dipole but can play a significant role when electric dipole transitions are forbidden.

The expression (4.40) is valid for a multi-electron atom, but in fact, the sum over i is a sum over electrons, and for the case of spontaneous emission, there is usually only one electron in the excited state of interest. Hence the only difference between the situation in a complex atom and the situation in hydrogen is the form of the wave functions for the upper and lower states. Because the wave functions for hydrogen are both known and available to you, further detailed development of these concepts will be done for hydrogen, but the expressions have a more general validity.

4.3 Lifetime for Hydrogen

Simplifying the problem to a one-electron system, Eq. (4.40) can be written as

$$w_{d\Omega} = \frac{\omega^3 e^2}{2\pi\hbar c^3}|\vec{r}_{m\ell} \cdot \vec{\epsilon}|^2 d\Omega \tag{4.41}$$

where $\vec{r}_{m\ell} \equiv \langle m|\vec{r}|\ell\rangle$.

It is worth mentioning here that this expression is a good deal more complicated that it might at first appear. The reason is that the operator is an inner product of two vectors whose value depends on the direction in which one is looking. The wave functions themselves also depend on spatial directions, and so the result becomes dependent on polarization (through $\vec{\epsilon}$), the direction in which the photon is emitted, and whether or not an ensemble of atoms has a statistical distribution of m substates. All of this is significantly beyond our present concerns but are mentioned to alert the reader to what can be found in more advanced treatments. If one doesn't care about the direction of the emitted photon, counts two for the polarization, and assumes statistical population of sublevels, a significant simplification results.

With these comments in mind, $w_{d\Omega}$ can be integrated over all directions to obtain the transition probability per unit time of a photon having any possible direction. This is equivalent to integrating (4.41) for all possible directions of $\vec{\epsilon}$ keeping $\vec{r}_{m\ell}$ fixed in space. Choose the z-axis of our coordinate frame to lie along $\vec{r}_{m\ell}$; then $\vec{r}_{m\ell} \cdot \vec{\epsilon} = |\vec{r}_{m\ell}| \cos\theta$ and

$$\int w_{d\Omega}\, d\Omega = \frac{\omega^3 e^2}{2\pi\hbar c^3}\, |\vec{r}_{m\ell}|^2 \int \cos^2\theta\, d\Omega$$

$$= \frac{2\omega^3 e^2}{3\hbar c^3}\, |\vec{r}_{m\ell}|^2$$

This would be sufficient if there was one polarization vector $\vec{\epsilon}$ or if it was guaranteed that all emitted radiation would be plane polarized. There are, however, two independent states of polarization, so multiply the above by 2. (Usually $\vec{\epsilon}$ is written $\epsilon^{(\alpha)}$ where $\alpha = 1$ or 2. This has been suppressed for legibility.) Finally, the total transition probability per unit time for one-electron systems is given by

$$w = \frac{4e^2\omega^3}{3\hbar c^3}\, |\langle m\,|\vec{r}|\,\ell\rangle|^2 \qquad (4.42)$$

This is fine but is still somewhat abstract. ℓ is the initial state and m is the final state and $\omega = (E_\ell - E_m)/\hbar$ and $E_\ell > E_m$. Since (4.42) holds for spontaneous emission of one-electron atoms it follows that

$$|\ell\rangle = |\, n_i\, \ell_i\, m_{\ell_i}\rangle \quad i \text{ initial}$$

$$\langle m| = \langle n_f\, \ell_f\, m_{\ell_f}\,| \quad f \text{ final}$$

Of course the ℓ and m on the LHS of the above equations have nothing to do with the ℓ and m on the RHS! For what follows, think only of n, ℓ, and m quantum numbers of hydrogenic systems.

In the absence of external fields or some other way to determine a direction in space, there will be m degeneracy. Despite the actual degeneracy in ℓ, the different values are nonetheless distinct, written, for example, as $3d$ and $3p$. They are indeed distinct entities though they happen to be energetically degenerate for this Hamiltonian.

Since the m values in some sense select spatial directions and all directions are equally probable, the usual prescription is to *average over initial states* and *sum over final states*. Then

$$w(n_i\ell_i \to n_f\ell_f) = \frac{4e^2\omega^3}{3\hbar c^3}\frac{1}{2\ell_i+1}\sum_{m_i\,m_f}|\langle n_f\,\ell_f\,m_f\,|\vec{r}|\,n_i\,\ell_i\,m_i\rangle|^2 \qquad (4.43)$$

So $|\langle\alpha_f\,|\vec{r}|\,\alpha_i\rangle|^2 = \sum_j\langle\alpha_f|x_j|\alpha_i\rangle\langle\alpha_i|x_j|\alpha_f\rangle$ where x_j is the jth Cartesian component of $\vec{r}$. There is a simple shortcut which leads to

$$|\langle\alpha_f\,|\vec{r}|\,\alpha_i\rangle|^2 \Rightarrow 3\,|\langle\alpha_f\,|z|\,\alpha_i\rangle|^2$$

There is no equality in general, but equality holds when the sum over all m_i and m_f is taken. Since $z = r\cos\theta$, the integral may be written as

$$\sum_{m_i\,m_f}|\langle n_f\,\ell_f\,m_f\,|r\cos\theta|\,n_i\,\ell_i\,m_i\rangle|^2 = \left|\int R_{n_i\ell_i}\,R_{n_f\ell_f}\,r^3\,dr\right|^2$$

$$\sum_{m_i\,m_f}\left|\int Y_{\ell_f}^{*m_f}(\theta,\phi)\cos\theta\,Y_{\ell_i}^{m_i}(\theta,\phi)\,d\Omega\right|^2 \qquad (4.44)$$

But

$$\cos\theta = \sqrt{\frac{4\pi}{3}}\,Y_1^0(\theta,\phi)$$

The above integral over Ω is nothing but $c^1(\ell_i m_i;\ell_f m_f)$ with $m_i = m_f$ (given by (2.20)) and zero otherwise. Alternatively one may use the expression for the integral over three spherical harmonics, Eq. (2.19), to obtain

$$\sum_{m_i\,m_f}\left|\int Y_{\ell_f}^{*m_f}\cos\theta Y_{\ell_i}^{m_i}\,d\Omega\right|^2 = \frac{4\pi}{3}\left[\frac{(2\ell_f+1)(2\ell_i+1)3}{4\pi}\right]$$

$$\begin{pmatrix}\ell_f & 1 & \ell_i\\ 0 & 0 & 0\end{pmatrix}^2 \sum_{m_i\,m_f}\begin{pmatrix}\ell_f & 1 & \ell_i\\ -m_f & 0 & m_i\end{pmatrix}^2 \qquad (4.45)$$

and the integral is zero unless

$$m_i = m_f \quad\text{and}\quad \ell_i = \ell_f \pm 1.$$

By invoking Eq. (1.54), the sums over m_i and m_f simply yield $1/3$. The other $3-j$ symbol can be looked up for the condition that $\ell_i = \ell_f + 1$ to give

$$\begin{pmatrix} \ell_f + 1 & \ell_f & 1 \\ 0 & 0 & 0 \end{pmatrix}^2 = \frac{(\ell_f + 1)}{(2\ell_f + 3)(2\ell_f + 1)}$$

The entire angular integral is seen to be $(\ell_f + 1)/3$. Starting with the opposite possibility, $\ell_i = \ell_f - 1$, one obtains $(\ell_i + 1)/3$ for the angular integral.

Problem 4.1

Derive the above expressions.

Completing the evaluation of Eq. (4.43) and remembering to multiply by three produces

$$w(n_i\ell_i \rightarrow n_f\ell_f) = \frac{4}{3}\frac{e^2\omega^3}{\hbar c^3}\frac{\ell_f + 1}{2\ell_i + 1}\left| \int R_{n_i\ell_i}R_{n_f\ell_f}r^3\,dr\right|^2 \qquad (4.46)$$

The lifetime of the excited level is just

$$t(n_i\ell_i) = \frac{1}{\sum_f w(n_i\ell_i \rightarrow n_f\ell_f)} \qquad (4.47)$$

Equation (4.46) above is in Gaussian units. To obtain a particularly convenient expression, note that $\omega/2\pi c = 1/\lambda$ where λ is the wavelength of the emitted light which is also a measure of the energy difference of the transition usually given as $1/\lambda$ expressed in cm^{-1}. With this in mind, working out the constants yields

$$\boxed{w(n_i\ell_i \rightarrow n_f\ell_f) = 2.0259 \cdot 10^{-6}\left(\frac{1}{\lambda}\right)^3\frac{\ell_f + 1}{2\ell_i + 1}\left| \int R_{n_i\ell_i}R_{n_f\ell_f}r^3\,dr\right|^2\frac{1}{r_B^2}\,\sec^{-1}}$$

$$(4.48)$$

where r_B is the Bohr radius. If the integral is done in atomic units, $r_B = 1$.

Problem 4.2

Find the lifetimes of the $2p$, $3p$, $4p$, $3d$, and $4d$ levels. Only *one* lower state will dominate the sum in (4.47), so use only that one.

4.4 Transition Moments for Complex Atoms

The derivations required to obtain the transition moment and lifetime for hydrogenic atoms are not so different for complex atoms. It is useful to have these expressions available and to consider their consequences for higher multipoles and for selection rules. If one uses the spherical harmonics as a template on which to build more complex operators, then one can manipulate the resultant expressions using angular momentum algebra and take advantage of the Wigner–Eckart theorem and elegant ways of combining operators. These techniques are part of more advanced treatments but the final expressions and the physics which they convey are accessible to us now so long as we are willing to forego the level of detail of the previous sections.

If one defines the jth electric multipole moment as

$$Q_j^m \equiv e \left[\frac{4\pi}{2j+1} \right]^{1/2} \sum_i r_i^j Y_j^m (\theta_i, \phi_i) \qquad (4.49)$$

then the electric dipole transition probability can be written as

$$w(\mathcal{E}1) = \frac{4k^3}{3\hbar} [\langle \alpha JM | Q_1^m | \alpha' J' M' \rangle]^2 . \qquad (4.50)$$

This is the multi-electron analog to Eq. (4.42). The integral over directions has been performed. The sum over electrons occurs in the definition of Q_1^m. α, J, and M represent the quantum numbers of the lower state, while the primed ones are those of the upper state. k is nothing but ω/c. The electric quadrupole transition moment can be written as

$$w(\mathcal{E}2) = \frac{k^5}{15\hbar} [\langle \alpha JM | Q_2^m | \alpha' J' M' \rangle]^2 . \qquad (4.51)$$

Problem 4.3

The electric dipole moment occurs in the units of $e\, a_0$ where e is the electric charge and a_0 is the Bohr radius or unit of distance in atomic units. Then the dipole transition probability will be given by $\frac{4k^3}{3\hbar} e^2 a_0^2$ multiplied by a matrix element that at most can be of the order of unity. Estimate the maximum value of the transition probability for a dipole transition in the visible spectral range (500 nm).

Problem 4.4

Repeat the above calculation for the electric quadrupole transition probability and show that the ratio of the electric dipole to electric quadrupole probabilities is 4.5×10^7 for a transition at 500 nm.

What is not remotely obvious from our treatment of electric multipole transitions is that it is also possible to have magnetic multipole transitions. Again one requires an additional time-dependent term in the Hamiltonian, but this time it cannot be motivated from the classical Hamiltonian. Instead, one should consider all of the terms that occur when solving the hydrogen atom relativistically. Following such considerations, one can define the magnetic multipole moment to be

$$M_j^m \equiv \mu_0 \left[\frac{4\pi}{2j+1} \right]^{1/2} \sum_i \vec{\nabla} [r_i^j Y_j^m (\theta_i, \phi_i)] \bullet \left[\left(\frac{2}{j+1} \right) \vec{\ell}_i + 2 \vec{s}_i \right] \qquad (4.52)$$

Here μ_0 is the Bohr magneton and the sum is over individual orbital and spin angular momenta. If this expression looks a *lot* more complicated than the electric multipole one, rest assured that it is. But one rarely calculates these moments. It is often enough to realize that they exist and give rise to transitions. It is useful to know the order of magnitude of these various multipoles which is the purpose of the suggested problems. The expression for the magnetic dipole transition probability is then given by

$$w(\mathcal{M}1) = \frac{4k^3}{3\hbar} \left[\langle \alpha JM | M_1^m | \alpha' J' M' \rangle \right]^2 . \tag{4.53}$$

Problem 4.5

Perform the calculation done above for the electric quadrupole transition probability for the magnetic dipole transition, and find the ratio of the electric quadrupole to magnetic dipole transition probabilities. These are usually considered to be of comparable magnitude.

4.5   Lifetimes, Selection Rules, and Oscillator Strengths

Note from Eq. (4.46) of the previous section that the transition probability varies as w^3. When considering order of magnitude estimates for lifetimes of excited states in which the transition frequency spans many decades, one can take (4.48) and write a crude estimate (which sets the matrix element to unity) as

$$t(n_i \rightarrow n_f) \approx \frac{10^{37}}{\nu^3} \text{ sec}$$

Using this it is possible to make a table of approximate lifetimes of excited states against *electric dipole radiation*.

	Radio	Microwave	IR	Visible	UV
ν (Hz)	10^3	10^{10}	10^{13}	5×10^{14}	5×10^{15}
t (s)	10^{28}	10^7	10^{-2}	3×10^{-6}	3×10^{-9}

Problem 4.6

Very high Rydberg states in hydrogen are receiving experimental attention today. For such states the Yrast chain ($n_i, \ell_i = n_i - 1; n_f = n_i - 1, \ell_f = n_f - 1$) often dominates. Estimate the lifetime of the $n = 50$ level in hydrogen against an Yrast transition.

The most straightforward way to derive selection rules is to return to (4.44) and to recall that

$$\int Y_{\ell_f}^{*m_f} Y_1^0 Y_{\ell_i}^{m_i} \, d\Omega \quad \text{is proportional to} \quad \begin{pmatrix} \ell_i & 1 & \ell_f \\ 0 & 0 & 0 \end{pmatrix}$$

This $3-j$ symbol is zero unless ℓ_i, ℓ_f, and 1 can "form a triangle" and unless $\ell_i + \ell_f + 1$ is an even integer. This results in the rule that $\ell_f = \ell_i \pm 1$. *No other electric dipole transitions are possible.*

Since the parity of a state is given by $(-1)^\ell$, states connected by electric dipole transitions must have opposite parity. What is not obvious is that these rules hold for complex atoms as well, though one might expect additional rules following from the coupling of angular momenta.

Since in a complex atom the transition operator is just $\vec{r}_i$, a one-electron operator, there exist nonvanishing matrix elements between different determinantal states only if the states differ by a single spin orbital. Since the operator is spin independent, there can be no spin flip during a transition which leads to the selection rule $\Delta S = 0$. The total angular momentum J, along with L and S, appears in a $6-j$ symbol whose symmetries lead to selection rules for these.

Here then are the selection rules for electric dipole transitions, given in order of strongest to weakest:

1. Parity change
2. $\Delta J = 0, \pm 1$ but not $0 \to 0$
3. Only one different orbital between configurations with $\Delta \ell = \pm 1$
4. $\Delta L = 0, \pm 1$ but not $0 \to 0$
5. $\Delta S = 0$

Numbers (1) and (2) are never violated. (Well, hardly ever; number (2) can be violated if hyperfine structure, interaction with the nuclear spin, is relevant.) Number (3) is as good as the one-electron spin orbital description, and so can be violated when configuration interaction is considered. Numbers (4) and 5 are not any better than the LS coupling approximation and break down whenever intermediate or j–j coupling is used as a descriptor.

Problem 4.7

Which of the following transitions are forbidden? Note that transitions are *always* written with the *lower* energy level on the *left*.

1. $2p^2 \, {}^3P_1 - 2p3d \, {}^3P_2$
2. $2s^2 \, {}^1S_0 - 2p^2 \, {}^1D_2$
3. $2p^3 \, {}^4S_{3/2} - 2p^2 3p \, {}^4S_{3/2}$
4. $2p^3 \, {}^2D_{3/2} - 2p^2 3s \, {}^2D_{5/2}$
5. $2p^3 \, {}^2P_{3/2} - 2s2p3s \, {}^2P_{1/2}$
6. $2p^2 3s \, {}^2D_{3/2} - 2p^2 3p \, {}^2F_{5/2}$
7. $2p^2 3s \, {}^2P_{1/2} - 2p^2 3p \, {}^2D_{5/2}$
8. $3d^2 \, {}^1G_4 - 3d4f \, {}^3G_5$
9. $3d^2 \, {}^3F_2 - 4p4f \, {}^3G_3$
10. $2p^3 \, {}^2D_{3/2} - 2s3d^2 \, {}^2D_{5/2}$
11. $3d^2 4s \, {}^2G_{7/2} - 3d^2 4f \, {}^2G_{9/2}$

Here are some concepts, many of which predate quantum mechanics. First return to (4.42) which can be rewritten as

$$w = \frac{4}{3}\frac{e^2\omega^3}{\hbar c^3}|\langle 1|\vec{r}|2\rangle|^2$$

where 1 is the lower (final in emission) state and 2 is the upper state. Now define the line strength to be

$$S \equiv e^2 \sum_{m_i\, m_f} |\langle 1|\vec{r}|2\rangle|^2$$

This is perfectly OK for a complex atom since $\sum_i \vec{r}_i$ has only *one* nonzero matrix element between determinantal states, so no ambiguity can arise. This is not to say that S is easy to evaluate for complex atoms; it isn't.

The probability of transition from state 2 to 1 is just the Einstein A coefficient and is given by

$$A_{21} = \frac{4}{3}\frac{\omega^3 S}{\hbar c^3 g_2} \qquad \text{(Gaussian)}$$

where g_2 is the statistical weight of level 2 and for a one-electron system is $(2\ell+1)$. (The above is just (4.43).)

The Einstein B coefficients are related to stimulated emission (B_{21}) and absorption (B_{12}) and satisfy

$$g_1 B_{12} = g_2 B_{21}$$

The stimulated emission coefficient, B_{21}, is related to A_{21} by

$$B_{21} = \frac{\pi c^3 A_{21}}{2\hbar\omega^3}$$

Finally the oscillator strength, f_{12}, which is dimensionless in any system of units, is given by

$$f_{12} \equiv \frac{2m\omega S}{3e^2\hbar g_1} \tag{4.54}$$

4.6   Absorption

If one reflects for a moment on Eqs. (4.29) and (4.30), it is obvious that our expressions for the transition probability are valid in absorption as well as in emission. It is true that one has to be a bit careful when counting degeneracies; are they for the upper or lower state? One comment should be made here. The

Table 4.1 Selection rules for some radiative transitions from level 1 to level 2

	Electric dipole E1	Magnetic dipole M1	Electric quadrupole E2
Parity	$\pm1 \rightarrow \mp1$	$\pm1 \rightarrow \pm1$	$\pm1 \rightarrow \pm1$
ΔJ	$0, \pm1$	$0, \pm1$	$0, \pm1, \pm2$
	$J_1 + J_2 \geq 1$	$J_1 + J_2 \geq 1$	$J_1 + J_2 \geq 2$
ΔM	$0, \pm1$	$0, \pm1$	$0, \pm1, \pm2$
$\Delta \ell$	±1	0	$0, \pm2$
		$(\Delta n = 0)$	$\ell_1 = 0 \nrightarrow \ell_2 = 0$
			$(\Delta m_\ell = 0, \pm1, \pm2)$
ΔS	0	0	0
	$(\Delta M_S = 0)$		$(\Delta M_S = 0)$
ΔL	$0, \pm1$	0	$0, \pm1, \pm2$
	$Ł_1 + L_2 \geq 1$		$Ł_1 + L_2 \geq 2$
	$(\Delta M_L = 0, \pm1)$		

concept of initial or final state is much less useful than that of upper or lower when considering transitions. Expression for the f-value or oscillator strength has gained wide acceptance and is used for both emission and absorption.

The perception that conventional emission and absorption spectroscopy were done by physicists in the 1930s and 1940s but have since been relegated to pat diagnostic work by chemists or engineers has long been widespread but simply isn't true. Astronomical research alone uses both techniques extensively, and the laboratory search for spectroscopic lines, be they from diffuse interstellar bands (DIBS) or Fe XIV, has been a rich source of intense research up to the present time. Furthermore, the formulae presented here are applicable to laser-based research so long as pulse powers are not extreme, $>10^{14}$ W.

Despite the theoretical similarities between the expressions for emission and absorption, the two techniques have several fundamental differences. Perhaps the most useful one is that a measurement of an absorption line is a measurement of the f-value for the transition multiplied by the number density of the absorbing state. Consider Beer's law for a beam of light of intensity I_ν passing through an absorbing medium with a path length of l where ν is the wavevector, $1/\lambda$ of the light:

$$I_\nu(l) = I_\nu(0) \, e^{-k_\nu l}$$

Typically absorption spectroscopy is performed by measuring the spectrum of the source with, $I_\nu(l)$, and without, $I_\nu(0)$, the sample or absorbing medium present. By taking the logarithm of the ratio of the spectra, one obtains the absorption coefficient, k_ν:

$$k_\nu = \frac{1}{l} \, ln\frac{I_\nu(0)}{I_\nu(l)}.$$

Clearly this depends on knowing the path length of the light through the sample. A plot of k_ν vs ν is then an absorption spectrum and if done with the signs as specified above looks identical to an emission spectrum; that is, the absorption

lines are plotted more positively than the background. If one integrates under the
absorption profile for a given line with the conditions that the line has not saturated,
there is no spontaneous emission present, and the line is well isolated from lines of
other transitions in the medium under study, then that integrated intensity is directly
proportional to the f-value:

$$\boxed{\int_{line} k_\nu \, d\nu = \frac{\pi e^2}{mc^2} N_1 f_{12}.} \tag{4.55}$$

This expression is in Gaussian units with ν in cm^{-1} and N_1 being the number
density in the lower state, the state from which absorption is occurring.

Problem 4.8

Equation (4.55) can be expressed in SI units by replacing e^2 by $e^2/4\pi\epsilon_0$ and remembering that ν
needs to be in Hz. Show that the expression becomes

$$\int_{line} k_\nu \, d\nu = \frac{e^2}{4\epsilon_0 mc} N_1 f_{12}.$$

Absorption spectroscopy as outlined above has traditionally been used for
molecules in the infrared spectral region. It is limited by the fact that one needs
a sufficient number of absorbers to discern a change in intensity. Commonly the
absorbing level is the ground state or perhaps a metastable state. Since the signal
scales with the optical path length, long path-length cells have been designed to
measure weak absorptions. The longest path lengths are observed in interstellar
clouds in which over 100 molecules have been identified by their absorption line
signatures.

Part II
Diatomic Molecules

Chapter 5
Electronic Structure of Diatomic Molecules

What makes the solution for the electronic motion in a molecule so much more difficult from that for an atom (which as we have seen is certainly not easy!) is that the problem, in general, is multicentered. In an atom, with a single nucleus, one may use spherical coordinates for the multielectron problem. The angular part of the problem is effectively solved by invoking spherical harmonics for the basis functions for each electron. In a molecule, spherical symmetry is broken, and one cannot even use the same quantum numbers that one could use for an atom. It is little wonder that most texts that do treat the quantum mechanics of molecules ignore the atomic underpinnings. One can hardly blame the authors; the methodology is fundamentally different.

What this chapter will try to show is that, while different, the fundamentals that were learned in the previous part of the text can and should be kept in mind when attempting to solve molecular problems. This is easier to do for diatomic molecules than more complicated ones and easier still for molecular hydrogen and its one-electron molecular ion. Because our aim is to understand the fundamentals, only diatomic molecules will be considered. Realize that molecules have multiple nuclei; these nuclei can vibrate and rotate and such motions occur at much lower energies than electron excitation. In fact, such excitations occur at ordinary, everyday temperatures and represent much, if not most, of what chemists study. If chemistry is the study of reactions and interactions of molecules, the forming and breaking of bonds, there is much that can be learned without appeal to quantum mechanics, but that is not our concern. Our interest here is to consider what quantum mechanics has to offer for the study of molecules and to try to present the fundamentals in a manner that will be of use to physicists in all subdisciplines.

Now it is tempting to say that if the electrons of a molecule are in their ground states, the study of such molecules is physical chemistry, while if the molecules are electronically excited, the study is chemical physics. Tempting but false. The worlds of molecular physics, chemical physics, and physical chemistry have so much overlap that I have never met anyone who attempted exclusive definitions. It might be fair to say that most physicists working with molecules work with

R.L. Brooks, *The Fundamentals of Atomic and Molecular Physics*, Undergraduate Lecture Notes in Physics, DOI 10.1007/978-1-4614-6678-9_5, © Springer Science+Business Media New York 2013

diatomics at least most of the time. I am sure there are those who would disagree but even those had to learn about diatomic molecules first. Diatomics manage to straddle the divide between atoms, whose nuclear motion is irrelevant, and complex molecules in which knowing the arrangement of the nuclei can be a challenge. Group theory is indispensable for the study of complex molecules (ones that are multicentered) but not necessary for the study of diatomic molecules. Hence we shall limit ourselves to the simplest of diatomic molecules but treat the motions of the electrons and nuclei quantum mechanically.

5.1   Hydrogen Molecular Ion

The simplest molecule that one can consider is formed by two protons separated by a distance R surrounded by a single electron. Since there is a net positive charge, the system is clearly a positive ion and represents the hydrogen molecular ion. Just as the hydrogen atom is the only problem in atomic physics that can be exactly solved, so the hydrogen molecular ion is the only problem in molecular physics which has an exact solution. Here the analogy stops. In atomic physics the solution for hydrogen forms the foundation for complex atoms. The exact solution for $H_2{}^+$ does *not* form the foundation for complicated molecules. Rather an approximate solution is used for more complicated molecules, and this approximate solution can be performed on $H_2{}^+$ and then compared for accuracy with the exact solution. Hence we shall not derive the exact solution[1] but rather use some of the results to discuss general properties of diatomic molecules.

5.1.1   Born–Oppenheimer Approximation

Let us clarify the quantum mechanical problem posed by molecules. The "solution" of such a problem must not only describe the wave function for the electrons but must also consider the motion of the nuclei. The following description is valid for all molecules, not just diatomic ones.

Let the positions of the nuclei be given by coordinates X_i where X represents (X, Y, Z) and i means the ith nucleus. Similarly x_j is the position of the jth electron. Then the time-independent Schrödinger equation for the system as a whole can be written as

$$\left[\sum_i -\frac{\hbar^2}{2M_i} \nabla_i^2 + \sum_j -\frac{\hbar^2}{2m_e} \nabla_j^2 + V(X_i, x_j) \right] \psi(X_i, x_j) = \epsilon \psi(X_i, x_j) \quad (5.1)$$

---

[1] See J. Slater, *Quantum Theory of Molecules and Solids*, Vol. 1, Appendix 1.

Here $V(X_i, x_j)$ is the electrostatic interaction between all pairs of particles, electrons and nuclei, and contains all necessary sums within it. A two-step approximation can be performed toward attempting a solution of (5.1). The first is to solve the equation

$$\left[\sum_j -\frac{\hbar^2}{2m_e} \nabla_j^2 + V(X_i, x_j) \right] u(X_i, x_j) = E(X_i)u(X_i, x_j) \qquad (5.2)$$

This is a Schrödinger equation with the nuclear kinetic energy omitted and represents a problem for fixed X_i (each i). Clearly the energy will then be a parametrically dependent function of X_i, and in principle the problem would have to be solved on a grid of points for X_i. The second step is to solve the Schrödinger equation

$$\left[\sum_i -\frac{\hbar^2}{2M_i} \nabla_i^2 + E(X_i) \right] v(X_i) = \epsilon v(X_i) \qquad (5.3)$$

This equation solves for the motion of the nuclei in a potential given by $E(X_i)$ from (5.2).

The Born–Oppenheimer theorem now states that ϵ from (5.3) is a good approximation to ϵ of (5.1) and furthermore that a good approximation for ψ of (5.1) is given by

$$\psi(X_i, x_j) = u(X_i, x_j)v(X_i) \qquad (5.4)$$

This separation of nuclear and electronic motions is made possible by the vast differences in the mass of an electron and even the lightest nucleus. In this chapter the electronic problem represented by (5.2) will be considered. For diatomic molecules, once the center of mass motion has been removed, X_i reduces to a single parameter, the internuclear separation. The energy solution is then a function $E(R)$ of internuclear separation. Because $E(R)$ is the potential for the nuclear motion (vibration and rotation), the set of curves $E(R)$ are called potential energy curves and a suitable graph the potential energy diagram. The nuclear motion will be taken up in the next chapter.

5.1.2 Molecular Orbitals for H_2^+

The H_2^+ problem is solved in spheroidal coordinates μ, λ, and ϕ (right handed in that order; details given in a subsequent section):

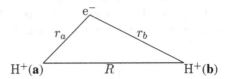

$$\mu \equiv \frac{r_a - r_b}{R} \qquad \lambda \equiv \frac{r_a + r_b}{R}$$

and ϕ is the angle of rotation about R. Note that $-1 \leq \mu \leq 1$ and $1 \leq \lambda \leq \infty$. The two nuclei, in this case just protons, are labeled **a** and **b** and are separated by a distance **R**. The vectors r_a and r_b then specify the position of the electron with respect to the two nuclei.

The Hamiltonian for electronic motion for H_2^+ is

$$\mathbf{H} = -\frac{\nabla^2}{2} - \frac{1}{r_a} - \frac{1}{r_b} + \frac{1}{R} \qquad (5.5)$$

Since $1/R$ is an effective constant, it can be dropped and added to the energy at the end. The wave function will be of the form

$$u = L(\lambda)M(\mu)e^{im\phi} \qquad (5.6)$$

The dependence of the solution on ϕ is like the atomic problem. However the z-axis of the problem is the internuclear axis, and this represents a defined direction, unlike the atomic problem. Hence the energy levels differ for differing $|m|$ but are doubly degenerate for $\pm m$. The levels (or more properly, curves, since each energy is a function of R) are labeled with a spectroscopic notation analogous to atoms. Levels with $m = 0$ are σ, $m = \pm 1$ are π, $m = \pm 2$ are δ, $m = \pm 3$ ϕ, etc., in analogy to s, p, d, f in atoms. This is a somewhat unfortunate analogy since in atoms s, p, d, f are labels for the ℓ quantum number, which does not exist in molecules since molecules are not even approximately spherically symmetric. Furthermore, for homonuclear molecules (two nuclei the same) the solutions either change sign on inversion (odd or ungerade) or they don't (even or gerade). This leads to σ_g, σ_u, π_g, π_u, etc. The solutions are then ordered in energy for small R and labeled (by Slater) in the order of appearance. So the first σ_g is $1\sigma_g$, the first σ_u is $1\sigma_u$, etc. This labeling is not standard and a modification will be introduced shortly. The only "good" quantum numbers for diatomics are m and parity. Clearly this makes labeling a bit of a challenge!

5.1.3 United Atom and Separated Atom Limits for H_2^+

Consider the lowest two potential energy curves for H_2^+, labeled $1\sigma_g$ and $1\sigma_u$. To a first approximation these curves can be formed by a linear combination of atomic orbitals (LCAO) in the following manner:

$$1\sigma_g \approx 1s_a + 1s_b = e^{-r_a} + e^{-r_b} \tag{5.7}$$

$$1\sigma_u \approx -1s_a + 1s_b = -e^{-r_a} + e^{-r_b} \tag{5.8}$$

where these orbitals are not normalized but should be. (To be discussed later.) In terms of spheroidal coordinates λ and μ the above can be expressed as

$$1\sigma_g \approx 2e^{-R\lambda/2} \cosh(R\mu/2) \tag{5.9}$$

$$1\sigma_u \approx 2e^{-R\lambda/2} \sinh(R\mu/2) \tag{5.10}$$

Problem 5.1

Derive (5.9) and (5.10) from (5.7) and (5.8).

Expressions like (5.9) and (5.10) are called molecular orbitals (MO). As $R \to 0$, the problem becomes equivalent to He^+ and it is expected that one of these orbitals will become the $1s$ orbital of He^+. The $1\sigma_g$ becomes this orbital, but note that our LCAO approximation approaches e^{-r} rather than e^{-2r} as it should. What happens to $1\sigma_u$? It becomes the $2p$ ($m = 0$) orbital of He^+. The limit $R = 0$ is called the united atom limit and thus offers an alternate scheme for labeling molecular orbitals. The $1\sigma_g$ could then be called the $1s\,\sigma_g$ and the $1\sigma_u$ the $2p\,\sigma_u$ where $1s$ and $2p$ are the $n\,\ell$ quantum numbers of the united atom.

In the united atom limit, the states are even or odd (g or u) depending on whether ℓ for the united atom is even or odd. For the separated atom, the molecular orbitals are even or odd depending on whether the separated atom orbitals have been added or subtracted. Now the molecule (as well as the separated atom) is a two-center problem which introduces a doubling to the number of orbitals. Nevertheless a one-to-one mapping between the united atom designation and the separated atom designation can be established once the notation is clarified.

United Atom Limit

In the united atom limit, the molecular orbitals become the atomic orbitals of He^+. Ignoring spin, there are the $1s$ ($m_\ell = 0$), $2s(0)$, $2p(-1)$, $2p(0)$, $2p(+1)$, $3s(0)$, etc. But the $2p(-1)$ and $2p(+1)$ are degenerate in an axial electric field such that the $2p(\pm1)$ will be referred to as $2p\,\pi_u$ (u, because $p = 1$ is odd). This π_u is doubly degenerate as are all orbitals except σ's. Therefore the following united atom orbitals can be listed: $1s\,\sigma_g$, $2s\,\sigma_g$, $2p\,\sigma_u$, $2p\,\pi_u$, $3s\,\sigma_g$, $3p\,\sigma_u$, $3p\,\pi_u$, $3d\,\sigma_g$, $3d\,\pi_g$, $3d\,\delta_g$, etc.

Separated Atom Limit

In the separated atom limit, the molecular orbitals become linear combinations of atomic orbitals of H. Don't be concerned about not having two electrons. The one electron could be on either nucleus (proton) and so forming these linear combinations is appropriate. Their occupation is another matter and for the problem being considered, H_2^+, only one electron will occupy any molecular orbital. (By adding and subtracting atomic orbitals, one can form g and u states out of any like ones.) To indicate the separated atom limit, the $n\,\ell$ designation will be written second. The molecular orbitals can then be written as follows: $\sigma_g\,1s$, $\sigma_u\,1s$, $\sigma_g\,2s$, $\sigma_u\,2s$, $\sigma_g\,2p$, $\sigma_u\,2p$, $\pi_g\,2p$, $\pi_u\,2p$, etc.

5.1.4   Variational Calculation of Ground-State MO for H_2^+

Let us calculate the molecular orbitals and energies for the lowest potential of the hydrogen molecular ion. Recall the previous comments that this solution will not be exact but rather will proceed in a manner that can be used for more complex molecules. This problem is ideally suited for working with prolate spheroidal coordinates—μ, λ, ϕ. This also affords an opportunity to practice techniques that all physics students learn but rarely exercise.

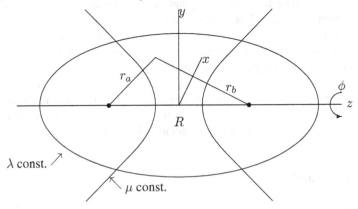

$$\lambda \equiv \frac{r_a + r_b}{R} \qquad \mu \equiv \frac{r_a - r_b}{R} \tag{5.11}$$

The transformation equations are given by

$$x = \frac{R}{2} \left[(\lambda^2 - 1)(1 - \mu^2) \right]^{1/2} \cos\phi \qquad (5.12)$$

$$y = \frac{R}{2} \left[(\lambda^2 - 1)(1 - \mu^2) \right]^{1/2} \sin\phi \qquad (5.13)$$

$$z = \frac{R}{2} \lambda\mu \qquad (5.14)$$

Problem 5.2

Show that the volume element $dV = dx\,dy\,dz$ becomes

$$dV = \frac{R^3}{8}(\lambda^2 - \mu^2)\,d\phi\,d\mu\,d\lambda \qquad (5.15)$$

Note that $-1 \le \mu \le 1, 1 \le \lambda \le \infty, 0 \le \phi \le 2\pi$, so the integral of a function over all space becomes

$$\int_{\text{all space}} f\,dV \Rightarrow \frac{R^3}{8} \int_0^{2\pi} d\phi \int_{-1}^1 d\mu \int_1^\infty f(\mu, \lambda, \phi)(\lambda^2 - \mu^2)\,d\lambda \qquad (5.16)$$

Previously the approximation was made that

$$1\sigma_g \approx e^{-r_a} + e^{-r_b} \qquad (5.7)$$

which is a simple linear combination of $1s$ orbitals on the separated atom, and it was found that the behavior as $R \to 0$ was particularly poor. The approximation can be improved by writing

$$1\sigma_g = N(e^{-\alpha r_a} + e^{-\alpha r_b}) \qquad (5.17)$$

$$= 2Ne^{-\alpha R\lambda/2} \cosh\left(\frac{\alpha R\mu}{2}\right) \qquad (5.17a)$$

where N is a normalization constant which needs to be evaluated and α is an adjustable parameter for each value of R, i.e., $\alpha = \alpha(R)$. From work in the previous sections, one would suspect that $\alpha(R = 0) = 2$ and $\alpha(R = \infty) = 1$. The task at hand is to minimize the total energy for a given R by varying α and thereby obtain a good ground-state orbital in the form of (5.17).

To find the normalization constant, set

$$\int (1\sigma_g)^2 \, dV = 1 \quad \text{or}$$

$$\frac{1}{N^2} = \int \left[e^{-2\alpha r_a} + e^{-2\alpha r_b} + 2e^{-\alpha(r_a+r_b)} \right] dV \tag{5.18}$$

Since the integration is being done over all space, it is permissible to perform the first two integrals using spherical coordinates centered on the appropriate nucleus. The first becomes

$$4\pi \int_0^\infty e^{-2\alpha r} r^2 \, dr \tag{5.19}$$

This is an easy integral to perform by parts. There is, however, a general equation for integrals of this type which will prove handy:

$$\int_y^\infty x^n e^{-ax} \, dx = \frac{n! e^{-ay}}{a^{n+1}} \left[1 + ay + \frac{(ay)^2}{2!} + \cdots + \frac{(ay)^n}{n!} \right] \tag{5.20}$$

It is easy to see that (5.19) becomes $8\pi/(2\alpha)^3 = \pi/\alpha^3$. Equation (5.18) now becomes

$$\frac{1}{N^2} = \frac{2\pi}{\alpha^3} + \int 2e^{-\alpha(r_a+r_b)} \, dV \tag{5.21}$$

Recognize that the integrand in (5.18) can be thought of as the electron probability density. The integrals which have been performed can be interpreted as the total probability of the electron being localized around either nucleus. The integral in (5.21) can be thought of as the overlap probability. More specifically, one usually defines the overlap integral as

$$S \equiv \frac{\alpha^3}{\pi} \int e^{-\alpha(r_a+r_b)} \, dV \tag{5.22}$$

from which one can see that

$$N = \left[\frac{\alpha^3}{2\pi(1+S)} \right]^{1/2} \tag{5.23}$$

Next the integration of (5.22) needs to be performed.

$$S = \frac{\alpha^3}{\pi} \int e^{-R\alpha\lambda} \, dV = \frac{\alpha^3 R^3}{4} \int_{-1}^1 d\mu \int_1^\infty e^{-R\alpha\lambda}(\lambda^2 - \mu^2) \, d\lambda$$

where (5.16) has been used. Now using (5.20) one obtains

$$S = \frac{\alpha^3 R^3}{4} \int_{-1}^{1} \left[\frac{2e^{-R\alpha}}{(R\alpha)^3} \left(1 + R\alpha + \frac{R^2\alpha^2}{2} \right) - \frac{\mu^2 e^{-R\alpha}}{R\alpha} \right] d\mu$$

and finally

$$S = e^{-R\alpha} \left(1 + R\alpha + \frac{R^2\alpha^2}{3} \right) \tag{5.24}$$

The normalized molecular orbital may be written

$$1\sigma_g = \left[\frac{\alpha^3}{2\pi(1+S)} \right]^{1/2} \left[e^{-\alpha r_a} + e^{-\alpha r_b} \right] \tag{5.25}$$

Using this MO the total energy of the molecule needs to be evaluated and then minimized by varying α.

The wave function in spheroidal coordinates is given by

$$\psi = 1\sigma_g = 2Ne^{-\alpha R\lambda/2} \cosh\left(\frac{\alpha R\mu}{2} \right) \tag{5.26}$$

$$\text{with} \quad N = \left[\frac{\alpha^3}{2\pi(1+S)} \right]^{1/2} \tag{5.23}$$

The total energy is

$$\int \psi^* \mathbf{H}\psi \, dV = \int \psi^* \left[-\frac{\nabla^2}{2} - \frac{1}{r_a} - \frac{1}{r_b} \right] \psi \, dV \tag{5.27}$$

First consider the kinetic energy. The Laplacian in spheroidal coordinates is

$$\nabla^2 \psi = \frac{4}{R^2} \left\{ \frac{1}{\lambda^2 - \mu^2} \frac{\partial}{\partial\lambda} \left[(\lambda^2 - 1)\frac{\partial\psi}{\partial\lambda} \right] \right.$$
$$\left. + \frac{1}{\lambda^2 - \mu^2} \frac{\partial}{\partial\mu} \left[(1 - \mu^2)\frac{\partial\psi}{\partial\mu} \right] + \frac{1}{(\lambda^2 - 1)(1 - \mu^2)} \frac{\partial^2\psi}{\partial\phi^2} \right\} \tag{5.28}$$

Obviously, from (5.26) $\frac{\partial\psi}{\partial\phi} = 0$.

$$\frac{\partial\psi}{\partial\lambda} = -\frac{\alpha R}{2}\psi \qquad\qquad \frac{\partial^2\psi}{\partial\lambda^2} = \frac{\alpha^2 R^2}{4}\psi$$

$$\frac{\partial\psi}{\partial\mu} = \frac{\alpha R}{2}\tanh\left(\frac{\alpha R\mu}{2} \right)\psi \qquad \frac{\partial^2\psi}{\partial\mu^2} = \frac{\alpha^2 R^2}{4}\psi$$

Using these expressions yields

$$\nabla^2\psi = \frac{(\lambda^2-1)\alpha^2}{\lambda^2-\mu^2}\psi + \frac{(1-\mu^2)\alpha^2}{\lambda^2-\mu^2}\psi - \frac{4\alpha\lambda}{R(\lambda^2-\mu^2)}\psi - \frac{4\alpha\mu}{R(\lambda^2-\mu^2)}\psi\tanh\left(\frac{\alpha R\mu}{2} \right)$$

The first and second terms readily sum, leaving

$$\nabla^2 \psi = \alpha^2 \psi - \frac{4\alpha\psi}{R(\lambda^2 - \mu^2)} \left[\lambda + \mu \tanh \left(\frac{\alpha R\mu}{2} \right) \right]$$

The kinetic energy is then

$$\int \psi^* \left(-\frac{\nabla^2}{2} \right) \psi \, dV = -\frac{\alpha^2}{2} \tag{5.29a}$$

$$+ \frac{\alpha^4 R^2}{1+S} \int_1^\infty \lambda e^{-\alpha R\lambda} \, d\lambda \int_{-1}^1 \cosh^2 \left(\frac{\alpha R\mu}{2} \right) d\mu \tag{5.29b}$$

$$+ \frac{\alpha^4 R^2}{1+S} \int_1^\infty e^{-\alpha R\lambda} \, d\lambda \int_{-1}^1 \frac{\mu}{2} \sinh(\alpha R\mu) \, d\mu \tag{5.29c}$$

(5.29b) yields

$$\frac{\alpha^4 R^2}{1+S} \left[\frac{e^{-\alpha R}(1+\alpha R)}{(\alpha R)^2} \right] \left[\frac{\sinh(\alpha R)}{\alpha R} + 1 \right]$$

(5.29c) yields

$$\frac{\alpha^4 R^2}{1+S} \left[\frac{e^{-\alpha R}}{\alpha R} \left(\frac{\cosh(\alpha R)}{\alpha R} - \frac{\sinh(\alpha R)}{(\alpha R)^2} \right) \right]$$

(5.29b) + (5.29c) gives

$$\frac{\alpha^2}{1+S} \left[1 + e^{-\alpha R}(1+\alpha R) \right]$$

Finally, the kinetic energy is given by (5.29a) + (5.29b) + (5.29c).

$$KE = \frac{1}{2} \frac{\alpha^2}{1+S} \left[1 + e^{-\alpha R} \left(1 + \alpha R - \frac{\alpha^2 R^2}{3} \right) \right] \tag{5.30}$$

Problem 5.3

Perform the necessary sums to obtain Eq. (5.30).

The potential energy operator is

$$-\frac{1}{r_a} - \frac{1}{r_b} = -\frac{r_a + r_b}{r_a r_b} = -\frac{4\lambda}{R(\lambda^2 - \mu^2)}$$

$$\int \psi^* PE \psi \, dV = -4\pi N^2 R^2 \int \lambda e^{-\alpha R\lambda} \, d\lambda \int \cosh^2 \left(\frac{\alpha R\mu}{2} \right) d\mu$$

$$= -\frac{2\alpha^3 R^2}{1+S} \int_1^\infty \lambda e^{-\alpha R\lambda} \, d\lambda \int_{-1}^1 \cosh^2 \left(\frac{\alpha R\mu}{2} \right) d\mu$$

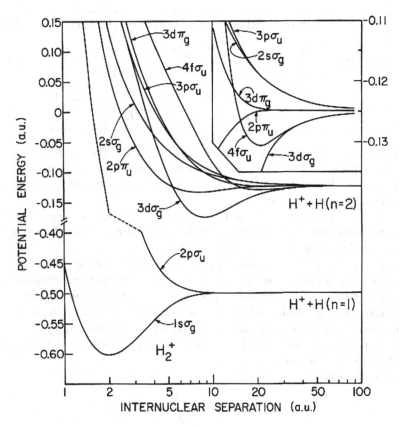

Fig. 5.1 Potential energy curves for H_2^+. Note how the curves approaching the atomic $n = 2$ asymptote evolve into Stark-split levels of H at 20 a.u.

This is the same integral as (5.29b) previously.

$$PE = -\frac{2\alpha}{1+S} \left[e^{-\alpha R}(1+\alpha R)\right] \left[\frac{\sinh(\alpha R)}{\alpha R} + 1\right]$$

$$PE = -\frac{\alpha}{1+S} \left[2e^{-\alpha R}(1+\alpha R) + \left(1 + \frac{1}{\alpha R}\right) - e^{-2\alpha R}\left(1 + \frac{1}{\alpha R}\right)\right] \quad (5.31)$$

The total energy is the sum of (5.30) and (5.31). This can be written out explicitly writing S from (5.24) in the following form:

$$\text{Total energy} = \text{KE} + \text{PE} = \alpha^2 F_1(\alpha R) + \alpha F_2(\alpha R) \qquad (5.32)$$

$$F_1(\alpha R) \equiv \frac{1}{2} \left[\frac{1 + e^{-\alpha R}(1 + \alpha R - \alpha^2 R^2/3)}{1 + e^{-\alpha R}(1 + \alpha R + \alpha^2 R^2/3)} \right] \qquad (5.33)$$

$$F_2(\alpha R) \equiv - \left[\frac{1 + \frac{1}{\alpha R} + 2e^{-\alpha R}(1 + \alpha R) - e^{-2\alpha R}(1 + \frac{1}{\alpha R})}{1 + e^{-\alpha R}(1 + \alpha R + \alpha^2 R^2/3)} \right] \qquad (5.34)$$

It has taken a lot of manipulation to get to this point. These expressions give the energy but are a function of the variational parameter α which still needs to be determined. Before doing that in general these expressions can be checked for the limits $R = 0$ and $R = \infty$.

At $R = \infty$, $F_1(\alpha R) = +\frac{1}{2}$ and $F_2(\alpha R) = -1$. Since α will equal 1 for the separated atom limit, the total energy will be $-\frac{1}{2}$ Hartree, correct for hydrogen, and the virial theorem[2] ($\text{KE} = -\frac{1}{2}\text{PE}$) is satisfied.

At $R = 0$, $F_1(\alpha R) = +\frac{1}{2}$ and $F_2(\alpha R) = -2$. To see this latter limit, it is not sufficient to replace the exponentials by the lead term of unity in an expansion about the origin. Rather $e^{-2\alpha R}$ needs to be replaced by $(1 - 2\alpha R)$ because the exponential is multiplied by $\frac{1}{\alpha R}$ which produces a constant term of -2. It makes no difference to the other terms whether the exponentials are replaced by unity or by $(1 - \alpha R)$. Further practice, in which even higher order terms must be kept, is provided by Problem 5.4 in the next section.

The total energy for $R = 0$ is then

$$E(R = 0) = \frac{\alpha^2}{2} - 2\alpha$$

Recall that α is a function of R and needs to be evaluated by minimizing the energy. This means

$$\frac{dE}{d\alpha} = 0 = \alpha - 2$$

so $\alpha = 2$ which is the expected value from our previous work. So the KE $= +2$ and PE $= -4$ for a total of -2 Hartrees, correct for He^+, and again the virial theorem is satisfied. Of course the above variational procedure at $R = \infty$ yields $\alpha = 1$. It is, however, *much* more difficult to evaluate α for intermediate values of R. This procedure was worked out before computers became ubiquitous but still has some merit. Let $\alpha R = w$, and then write

$$E = \alpha^2 F_1(w) + \alpha F_2(w)$$

$$\frac{dE}{d\alpha} = 0 = 2\alpha F_1 + \alpha^2 R \frac{dF_1}{dw} + F_2 + \alpha R \frac{dF_2}{dw}$$

---

[2]Valid for inverse square force *only*.

Fig. 5.2 α vs R for $H_2{}^+$

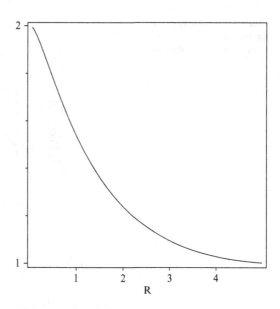

and

$$\alpha = -\frac{F_2 + w\frac{\mathrm{d}F_2}{\mathrm{d}w}}{2F_1 + w\frac{\mathrm{d}F_1}{\mathrm{d}w}}$$

What is wanted is α for a selection of values of R. Choosing instead a selection of values for w, one can use the above equation to find α and then $R = w/\alpha$. On first reading this may seem circuitous or even silly, but it's a clever way to solve an otherwise formidable expression for α. The solution for $\alpha(R)$ is shown in Fig. 5.2. Figure 5.3 compares three different solutions for the energy. The one marked "variational" represents the solution given here. The one marked "LCAO," for linear combination of atomic orbitals, is the solution one obtains using an α of unity. The "exact" solution is taken from Sharp.[3]

5.1.5   Variational Calculation of First Excited State MO for $H_2{}^+$

What about the solution for $1\sigma_u$, the first excited state for $H_2{}^+$? That molecular orbital can be written as

---

[3]T.E. Sharp, Atomic Data **2**, 119 (1971).

Fig. 5.3 Total energy (less
nuclear repulsion) for $H_2{}^+$
from different calculations

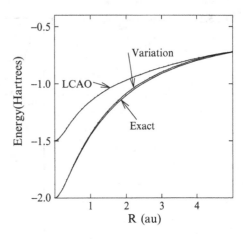

$$1\sigma_u = N\left(-e^{-\alpha r_a} + e^{-\alpha r_b}\right)$$

$$\text{where} \quad N = \left[\frac{\alpha^3}{2\pi(1-S)}\right]^{1/2}$$

with S as defined previously. In terms of prolate spheroidal coordinates, the
molecular orbital is

$$1\sigma_u = 2Ne^{-\lambda R\alpha/2}\sinh\left(\frac{\alpha R\mu}{2}\right)$$

The derivation proceeds as before, but now the kinetic energy is

$$\text{KE} = \frac{1}{2}\frac{\alpha^2}{1-S}\left[1 - e^{-\alpha R}\left(1 + \alpha R - \frac{\alpha^2 R^2}{3}\right)\right]$$

and the potential energy is

$$\text{PE} = -\frac{\alpha}{1-S}\left[1 + \frac{1}{\alpha R} - 2e^{-\alpha R}(1 + \alpha R) - e^{-2\alpha R}\left(1 + \frac{1}{\alpha R}\right)\right]$$

Writing the total energy as before in terms of the auxiliary functions F_1 and F_2,
which are given by

$$F_1(\alpha R) = \frac{1}{2}\left[\frac{1 - e^{-\alpha R}(1 + \alpha R - \alpha^2 R^2/3)}{1 - e^{-\alpha R}(1 + \alpha R + \alpha^2 R^2/3)}\right]$$

$$F_2(\alpha R) = -\left[\frac{1 + \frac{1}{\alpha R} - 2e^{-\alpha R}(1 + \alpha R) - e^{-2\alpha R}(1 + \frac{1}{\alpha R})}{1 - e^{-\alpha R}(1 + \alpha R + \alpha^2 R^2/3)}\right]$$

the total energy is then written as

$$E(\alpha R) = KE + PE = \alpha^2 F_1(\alpha R) + \alpha F_2(\alpha R).$$

As before, when $R = \infty$, $F_1(\alpha R = \infty) = +\frac{1}{2}$ and $F_2(\alpha R = \infty) = -1$. By setting $\frac{dE}{d\alpha} = 0$, one obtains $\alpha = 1$ and the energy is $-\frac{1}{2}$ Hartree, appropriate for the ground state of neutral H. What about for $R = 0$?

Problem 5.4

Show that $F_1(\alpha R = 0) = \frac{5}{2}$ and $F_2(\alpha R = 0) = -2$.

So $E(R = 0) = \frac{5}{2}\alpha^2 - 2\alpha$

$$\frac{dE}{d\alpha} = 5\alpha - 2 = 0 \quad \text{and} \quad \alpha = \frac{2}{5}$$

Then $KE = +\frac{2}{5}$ and $PE = -\frac{4}{5}$, and the virial theorem is satisfied.

The total energy is -0.4 Hartree rather than the "correct" answer of -0.5 Hartree, which is the value suitable for the $n = 2$ state of He^+. Note that no real value of α can produce an energy of -0.5 Hartree.

There is no mystery here. The wonder is that it turns out so well when you consider that the variational function $1\sigma_u$ misbehaves at $R = 0$, i.e., it does not have a form closely approximating the $2p\,m_\ell = 0$ wave function of He^+, at least not close enough to get really good agreement.

5.2 The Hydrogen Molecule, H_2

The problem of the hydrogen molecule is, with respect to wave functions, very similar to H_2^+. Because there are two electrons, however, two complications arise. The first is that the Hamiltonian is more involved, leading to very much more complicated integral expressions for the total energy. The second is that the wave function must be a properly antisymmetrized Slater determinant. Let's consider the wave function problem first.

Write the atomic orbitals as

$$a \equiv \left(\frac{\alpha^3}{\pi}\right)^{1/2} e^{-\alpha r_a} \tag{5.35}$$

$$b \equiv \left(\frac{\alpha^3}{\pi}\right)^{1/2} e^{-\alpha r_b} \tag{5.36}$$

and the ground-state molecular orbital as

$$g \equiv [2(1+S)]^{-\frac{1}{2}} (a+b) \tag{5.37}$$

Comparison of the expressions with our previous expressions for H_2^+ shows that *they are the same.* $g = 1\sigma_g$ previously and S is given by (5.24) of the previous section. Next, in complete analogy with the atomic problem, multiply this MO by a spinor. Call the spin-up spinor $\binom{1}{0} \equiv \gamma$ and the spin-down spinor $\binom{0}{1} \equiv \delta$. (These are usually labeled α and β, but α has been defined as the variational parameter.) The state function (for the two-electron problem) can be formed as

$$\psi(1,2) = \frac{1}{\sqrt{2}} \begin{vmatrix} g(1)\gamma(1) & g(2)\gamma(2) \\ g(1)\delta(1) & g(2)\delta(2) \end{vmatrix}$$

or

$$\psi(1,2) = 2^{-\frac{1}{2}} g(1)g(2) [\gamma(1)\delta(2) - \delta(1)\gamma(2)] \tag{5.38}$$

1 and 2 label the electrons. This expression is completely analogous to the ground state of He in which the configuration is $1s^2$. g plays the role of $1s$ and the spinors are the same. Clearly the spinor part of (5.38) says that the electron spins are antiparallel or that the state $\psi(1,2)$ is a singlet state. Furthermore, if the problem is limited to a single molecular orbital such that (5.37) is the *only* MO under consideration, the state $\psi(1,2)$ is unique. (This is analogous to there being only one state for a closed shell atomic system.) Recall that any time the energy level under consideration is composed of a single state of a single Slater determinant, the matrix elements of one- or two-electron operators are given by considering just the principal diagonal. Since the Hamiltonian being considered is spin independent, the matrix elements of **H**, evaluated with respect to $\psi(1,2)$, are the same if $\psi(1,2)$ were given by

$$\psi(1,2) = 2^{-\frac{1}{2}} g(1)g(2) \tag{5.38a}$$

Next notice that

$$g(1)g(2) = \frac{1}{2(1+S)} [a(1)a(2) + a(1)b(2) + b(1)a(2) + b(1)b(2)] \tag{5.39}$$

(The analogous expression for $1s^2 \, {}^1S$ is just $1s(1)1s(2)$. (5.39) is complicated by being a two-center problem.)

The Hamiltonian for the two-electron, two-center problem being considered is given by

$$\mathbf{H} = -\frac{\nabla_1^2}{2} - \frac{\nabla_2^2}{2} - \frac{1}{r_{1a}} - \frac{1}{r_{2a}} - \frac{1}{r_{1b}} - \frac{1}{r_{2b}} + \frac{1}{r_{12}} + \frac{1}{R} \tag{5.40}$$

The total energy can be found by evaluating

$$E(\alpha, R) = \int \psi^*(1, 2)\mathbf{H}\psi(1, 2)\, dV \tag{5.41}$$

for any value of R. α is the variational parameter and can be found as before by setting $\frac{dE}{d\alpha} = 0$.

Clearly (5.41) becomes a formidable expression with the Hamiltonian of (5.40). However, all of the required integrals are tabulated in Table 5.1 taken from Slater.[4] (Each of the integrals could be done using methods similar to H_2^+. Only the $\frac{1}{r_{12}}$ term requires something new.)

The results of this calculation, while reasonable, are not nearly so good as for H_2^+. In particular, the energy at $R = \infty$ is 0.29 Hartree too high. What is interesting is that simplification of the wave function will improve the result! If instead of (5.39) the following equation were used

$$\psi(1, 2) = \frac{1}{\sqrt{2}}g(1)g(2) = N\left[a(1)b(2) + b(1)a(2)\right] \tag{5.42}$$

the behavior at $R = \infty$ becomes correct. Such a calculation, without varying α, was performed by Heitler and London. The variational problem, which is of interest here, was performed by Wang and Rosen.

Why does (5.42) work better than (5.39)? The terms $a(1)a(2)$ and $b(1)b(2)$ represent charge densities in which both electrons are on the same nucleus. While the contributions exist and should be included, the wave function of (5.39) grossly overestimates the effect. If one insists on trying to represent the ground-state potential energy curve with a single MO (minimal basis set), it is better to exclude such terms entirely and use (5.42).

Proceeding with such a calculation, remembering that the first item of business is to find the normalization constant N, the kinetic and potential energies can be expressed as

$$\text{KE} = \frac{\alpha^2}{1 + S^2}(1 - 2SK - S^2) \tag{5.43}$$

$$\text{PE} = \frac{\alpha}{1 + S^2}(-2 + 2J + 4KS + J' + K') + \frac{1}{R} \tag{5.44}$$

(S is the same as used for H_2^+.) Here J, K, J', etc., are the values of particular integrals given in Table 5.1.

---

[4]*Quantum Theory of Matter*, 2nd ed. John C. Slater, McGraw-Hill, N.Y., 1968.

Table 5.1 Integrals needed for the energy calculation of the hydrogen molecule

$\int a(1)(-\frac{1}{2}\nabla_1^2)a(1)\,dv_1 = \frac{\alpha^2}{2}$

$\int a^2(1)\left(\frac{-1}{r_{1a}}\right)\,dv_1 = -\alpha$

$\int a^2(1)a^2(2)\left(\frac{1}{r_{12}}\right)\,dv_1\,dv_2 = \frac{5}{8}\alpha$

$\int a(1)b(1)\,dv_1 = S = \exp(-w)(1+w+\frac{w^2}{3}) = 1 - \frac{1}{6}w^2 + \frac{1}{24}w^4 - \dots$

$\int a^2(1)\left(\frac{-1}{r_{1b}}\right)\,dv_1 = \alpha J = \alpha[-\frac{1}{w} + \exp(-2w)(1+\frac{1}{w})]$

$\qquad\qquad = \alpha(-1 + \frac{2}{3}w^2 - \frac{2}{3}w^3 + \frac{2}{5}w^4 - \dots)$

$\int a(1)b(1)\left(\frac{-1}{r_{1b}}\right)\,dv_1 = \alpha K = -\alpha\exp(-w)(1+w)$

$\qquad\qquad = \alpha(-1 + \frac{1}{2}w^2 - \frac{1}{3}w^3 + \frac{1}{8}w^4 - \dots)$

$\int a(1)(-\frac{1}{2}\nabla_1^2)b(1)\,dv_1 = -\alpha^2(K+S/2) = \alpha^2\exp(-w)(\frac{1}{2}+\frac{1}{2}w - \frac{1}{6}w^2)$

$\qquad\qquad = \alpha^2(\frac{1}{2} - \frac{5}{12}w^2 + \frac{1}{3}w^3 - \frac{1}{16}w^4 + \dots)$

$\int a^2(1)b^2(2)\left(\frac{1}{r_{12}}\right)\,dv_1\,dv_2 = \langle ab|g|ab\rangle = \alpha J'$

$\qquad\qquad = \alpha\left[\frac{1}{w} - \exp(-2w)\left(\frac{1}{w} + \frac{11}{8} + \frac{3}{4}w + \frac{1}{6}w^2\right)\right]$

$\qquad\qquad = \alpha(\frac{5}{8} - \frac{1}{12}w^2 + \frac{1}{60}w^4 - \dots)$

$\int a^2(1)a(2)b(2)\left(\frac{1}{r_{12}}\right)\,dv_1\,dv_2 = \langle aa|g|ab\rangle = \alpha L$

$\qquad\qquad = \alpha\left[\exp(-w)\left(w + \frac{1}{8} + \frac{5}{16w}\right) + \exp(-3w)\left(-\frac{1}{8} - \frac{5}{16w}\right)\right]$

$\qquad\qquad = \alpha(\frac{5}{8} - \frac{7}{48}w^2 + \frac{3}{64}w^4 - \dots)$

$\int a(1)b(1)b(2)a(2)\left(\frac{1}{r_{12}}\right)\,dv_1\,dv_2 = \langle ab|g|ba\rangle = \alpha K'$

$\qquad\qquad = \frac{1}{5}\alpha\left[-\exp(-2w)(-\frac{25}{8} + \frac{23}{4}w + 3w^2 + \frac{1}{3}w^3) + W'\right]$

$\qquad\qquad = \alpha[\frac{5}{8} - \frac{1}{4}w^2 + (\frac{3}{100} + \frac{4}{75}\ln 4)w^4 - \dots]$

where $W' = +\frac{6}{w}[S^2(C+\ln w) + S'^2 Ei(-4w) - 2SS' Ei(-2w)]$

and $S' = \exp(w)(1 - w + \frac{1}{3}w^2) = 1 - \frac{1}{6}w^2 + \frac{1}{24}w^4 - \dots$

$C = $ Euler's constant $\int_0^1 \frac{1-\exp(-t)}{t}\,dt - \int_1^\infty \frac{\exp(-t)}{t}\,dt = 0.57722$

$Ei(x) = $ integral logarithm, $Ei(-x) = -\int_x^\infty \frac{\exp(-t)}{t}\,dt$, where $x > 0$

The atomic orbitals, $a(1)$, $b(1)$, are defined in Eqs. (5.35) and (5.36). α is a variational parameter; energies are in Hartrees, distances in au, and $w = \alpha R$, where R is the internuclear separation

---

Problem 5.5

Using the values of Table 5.1, derive (5.43) and (5.44). Show that at $R = \infty$, the total energy is -1 Hartree, the sum of two hydrogen atoms in the $1s$ orbital.

The previous problem examined the behavior at $R = \infty$ and found that the energies were correct for two separated hydrogen atoms which is promising. But what happens in the opposite limit of $R = 0$, which would correspond to the ground state of helium? With no difficulty the energy as a function of α is given by

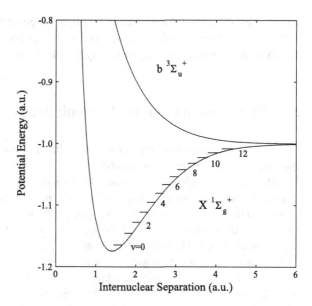

Fig. 5.4 Lowest two potential energy curves for H_2 showing positions of vibrational levels

$$E(\alpha, R = 0) = \alpha^2 - \frac{27}{8}\alpha \qquad (5.45)$$

$$\frac{dE}{d\alpha} = 0 = 2\alpha - \frac{27}{8}; \qquad \alpha = \frac{27}{16}$$

The second line evaluated the derivative which then yields $\alpha = {}^{27}\!/_{16}$ which gives the final result

$$E = -\left(\frac{27}{16}\right)^2 = -2.8477 \text{ Hartree}$$

The Hartree–Fock ground-state energy of helium is -2.8617 Hartree, so this approximation is not bad at least at $R = 0$ and $R = \infty$.

Now the equilibrium separation is 1.4 a.u. Wang reports an energy for this calculation of -1.139 Hartrees at equilibrium. The experimental value is -1.174. The difference is 0.035 Hartree. This must be considered excellent when one realizes that for He in the ground state, the difference between the previously quoted Hartree–Fock and the experimental value is 0.042 Hartree.

This energy difference between the HF and the "correct" value is known as the correlation energy. Configuration interaction is able to account for most of it. If the ground state of He is multi-configured with 11 configurations, Froese Fischer reports that ΔE reduces to 0.0007 Hartree.[5] Figure 5.4 shows the lowest two

---

[5]C. Froese Fischer, *The Hartree–Fock Method for Atoms: A Numerical Approach*, Wiley, N.Y., 1977.

potential energy curves for molecular hydrogen as given by Sharp.[6] The placement of the lowest 12 (of 14) vibrational energy levels has been included. The asymptote occurs at the sum of the energy of two hydrogen atoms.

5.3   State Designations for Diatomic Molecules

Even as $\ell_{\mathbf{z}}$ commutes with the one-electron diatomic molecule Hamiltonian, such that m_ℓ is a good quantum number, so $\mathbf{L}_z = \sum_i (\ell_{\mathbf{z}})_i$ commutes with the Hamiltonian of a multielectron molecule. This means that M_L, the component of angular momentum along the internuclear axis, is a constant of the motion.

For atoms, the designation of ℓ for a one-electron system is written as as s, p, d, etc., but the designations for L use the uppercase letters S, P, D, etc. For molecules the designations for M_L will be written as Σ, Π, Δ, etc., for multielectron diatomic molecules even as the lowercase σ, π, and δ were previously used for m_ℓ. Σ means $M_L = 0$ and is not degenerate, while $M_L = \pm 1 \Rightarrow \Pi$ is twofold degenerate and likewise for all larger values of M_L. $\mathbf{S}_z$ may be quantized along the same internuclear axis allowing the total S to be written as a superscript on the designation in analogy with atoms. Such a designation may look as $^1\Sigma$, $^3\Sigma$, etc.

Homonuclear (actually, equal charge) molecules have inversion symmetry. If the electronic wave function is unchanged under inversion, it is even (gerade); if it changes sign it is odd (ungerade). The easiest way to see this for the present case is to note that an inversion interchanges the nuclei so a state like $a(1)b(2) + a(2)b(1)$ is unchanged when a and b are reversed and must be a g state. The ground state for H_2, considered in the previous section, is then $^1\Sigma_g$. One caution here is that while it is true that an inversion exchanges identical nuclei, it is not true that an exchange of nuclei is the same as an inversion. This point is taken up in more detail in the following chapter.

Recall that the ground-state wave function for H_2 was originally written as $\psi(1,2) = Ng(1)g(2)$, which could have been written $(1\sigma_g)^2$. Even though these wave functions were truncated for a better approximation, this did not alter the symmetry, and these molecular orbitals afford a convenient labeling scheme. The ground state of H_2 could be called $(1\sigma_g)^2\,^1\Sigma_g$, while an excited orbital might be $1\sigma_g 1\sigma_u\,^1\Sigma_u$ or $^3\Sigma_u$. Of course a state like $(1\sigma_u)^2$ must be $^1\Sigma_g$.

There is one further symmetry which leads to a distinction for Σ *states only*. This is written as Σ^+ or Σ^- depending on whether the wave function changes sign when ϕ is replaced by $-\phi$. This will be introduced in the next section, but two points should be remembered:

1. The designation is for Σ states only.
2. The designation holds for all diatomic molecules, not just homonuclear ones.

---

[6]T.E. Sharp, Atomic Data **2**, 119 (1971).

5.4 First-Row Diatomic Molecules

Consider homonuclear diatomic molecules formed from like elements in the first row of the periodic table, i.e., Li_2 to F_2. Further consider that the molecular orbitals are formed from only the $\sigma\,1s$, $\sigma\,2s$, $\sigma\,2p$, and $\pi\,2p$ atomic orbitals on the separated atoms. By adding and subtracting these orbitals in like pairs, eight MOs can be formed, but the ordering of these MOs energetically is not obvious. It is found that they take the order

$$1\sigma_g \quad 1\sigma_u \quad 2\sigma_g \quad 2\sigma_u \quad 1\pi_u \quad 3\sigma_g \quad 1\pi_g \quad 3\sigma_u$$

Each σ orbital can accept two electrons (spin up and spin down), while each π orbital is doubly degenerate and can therefore accept four.

If these orbitals are filled in the order given above, then, for example, N_2, which has 14 electrons, would have the orbital designation

$$N_2: (1\sigma_g)^2(1\sigma_u)^2(2\sigma_g)^2(2\sigma_u)^2(1\pi_u)^4(3\sigma_g)^2 \; {}^1\Sigma_g$$

and there would be the equivalent of a closed subshell. What about O_2? O_2 has 16 electrons and it would have

$$O_2: \ldots \text{(as above)}(3\sigma_g)^2(1\pi_g)^2$$

but now the last orbital is only half full, so there must be some multiplet structure. Let's make a table of allowed states analogous to what was done for the atomic case. Next it is possible to infer the multiplet structure by examining the table in a

#	M_L	M_S	π_{g+}^{+}	π_{g+}^{-}	π_{g-}^{+}	π_{g-}^{-}
1	2	0	1	1	0	0
2	0	1	1	0	1	0
3	0	0	1	0	0	1
4	0	0	0	1	1	0
5	0	-1	0	1	0	1
6	-2	0	0	0	1	1

way similar to what was done for atoms. There are $\frac{4!}{2!2!} = 6$ ways of putting two particles into four boxes. The superscript on π_g represents a spin of $\pm^1/_2$, while the subscript represents $m_\ell = \pm 1$. The order of the columns is unimportant. What is important is that summing across the rows yields the total values for M_L and M_S. Now $M_L = \pm 2$ is a Δ state, and since $M_S = 0$ for the only two such occurrences and both orbitals are even, these two must have a ${}^1\Delta_g$ identity. The remaining four states all have $M_L = 0$, so clearly three of them belong to ${}^3\Sigma_g$, while one is ${}^1\Sigma_g$.

Just as in the atomic case, each line on the table above represents a Slater determinant, e.g., state #2 could be written

$$\phi_2 = \mathcal{A}\left(\pi_{g+}(1)\alpha(1)\,\pi_{g-}(2)\alpha(2)\right)$$

where α is the spin-up spinor and β will be the spin-down spinor. Writing out this wave function yields

$$\phi_2 = \frac{1}{\sqrt{2}}\left[\pi_{g+}(1)\pi_{g-}(2) - \pi_{g-}(1)\pi_{g+}(2)\right]\alpha(1)\alpha(2) \qquad (5.46)$$

Similarly ϕ_5 is given by

$$\phi_5 = \frac{1}{\sqrt{2}}\left[\pi_{g+}(1)\pi_{g-}(2) - \pi_{g-}(1)\pi_{g+}(2)\right]\beta(1)\beta(2) \qquad (5.47)$$

Clearly these wave functions belong to the triplet as their spins are aligned. So ϕ_3 and ϕ_4 must contribute to the singlet and to the $M_S = 0$ component of the triplet. These triplet and singlet components are linear combinations of ϕ_3 and ϕ_4. One such linear combination, which will be seen to belong to the triplet, is given by

$$
\begin{aligned}
\phi^t &= \frac{1}{\sqrt{2}}(\phi_3 + \phi_4) \\
&= \frac{1}{\sqrt{2}}\left[\pi_{g+}(1)\pi_{g-}(2) - \pi_{g-}(1)\pi_{g+}(2)\right]\frac{1}{\sqrt{2}}\left(\alpha(1)\beta(2) + \beta(1)\alpha(2)\right) \quad (5.48)
\end{aligned}
$$

The other linear combination is

$$
\begin{aligned}
\phi^s &= \frac{1}{\sqrt{2}}(\phi_3 - \phi_4) \\
&= \frac{1}{\sqrt{2}}\left[\pi_{g+}(1)\pi_{g-}(2) + \pi_{g-}(1)\pi_{g+}(2)\right]\frac{1}{\sqrt{2}}\left(\alpha(1)\beta(2) - \beta(1)\alpha(2)\right) \quad (5.49)
\end{aligned}
$$

There are two reasons why (5.48) belongs to the triplet. The first is that the spatial part of the function is the same as (5.46) and (5.47) and one should expect the three states belonging to $M_S = -1, 0, 1$ to differ only in their spin parts. The second is that the spin parts of (5.46) and (5.47) are symmetric under exchange of electrons and so is the spin part of (5.48). This symmetry must be the same for all components of a multiplet.

In the previous section it was mentioned that the symmetry operation $\phi \to -\phi$ afforded an additional designation for Σ states. Since all wave functions for diatomic molecules have some term like $e^{-iM_L\phi}$, if ϕ changes sign, it is equivalent to M changing sign so long as $M \neq 0$. Such an operation simply carries one degenerate state into the other for all but Σ states. Consider the Σ wave functions and make the substitution $\pi_{g+} \to \pi_{g-}$, by changing the sign of m_ℓ, and notice that the triplets change sign while the singlet does not. Hence the $^3\Sigma_g$ must be $^3\Sigma_g{}^-$, while the $^1\Sigma_g$ is $^1\Sigma_g{}^+$.

Hund's rule is still applicable for molecules, and this leads to the conclusion that the $^3\Sigma_g{}^-$ is the lowest orbital for O_2. This yields the interesting complication that O_2 is paramagnetic. Liquid O_2 when poured between the pole faces of a magnet is attracted like iron filings!

5.5 Bonding and Antibonding Orbitals

Look at a correlation diagram for homonuclear diatomic molecules. The left-hand side represents the united atom limit, while the right-hand side represents the separated atom limit. Consider only those molecular orbitals relevant to the first two rows (up to $Z = 10$) of the periodic table.

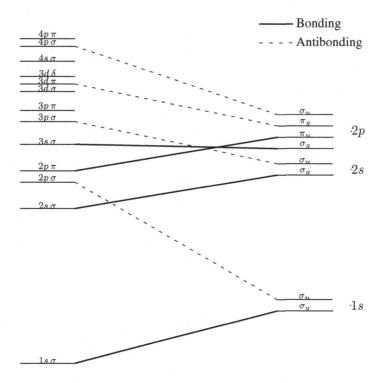

σ_g and π_u are bonding orbitals, while σ_u and π_g are antibonding.

Count $+1$ for every electron *pair* in a bonding orbital and -1 for every *pair* in an antibonding orbital. Count $\pm 1/2$ for a single electron in such an orbital. If the sum of such a count is positive, the molecule is bound. If zero or negative, it is unbound.

Consider three examples from first- and second-row elements:

1. He_2: Count is 0; not bound.
2. N_2: Count is $+3$.
3. O_2: Count is $+2$.

This indicates that nitrogen is more strongly bound than oxygen. Indeed N_2 is bound by 9.7 eV, while O_2 is bound by 5.1 eV.

While such an activity may be useful for developing an intuition for the chemical bond, it really is a poor substitute for doing the relevant quantum mechanics to determine whether and by how much an electron is bound to a molecule. This chapter has given you some idea of how such calculations can be performed even if the detail was far too sketchy to allow you to do them.

Chapter 6
Vibrations and Rotations of Diatomic Molecules

With the electronic part of the problem treated in the previous chapter, the nuclear motion shall occupy our attention in this one. In many ways the motion of two nuclei in a potential well formed by the electron cloud is among the simpler of all quantum mechanical problems. The reason is that it readily reduces to the motion of a single particle in a potential well. The derivation of how that comes about is given in detail in case some have not seen it before.

A single particle in a potential well is also what the hydrogen atom is, and it is useful to reflect on the differences. For hydrogen, one is interested in the motion of the electron in the coulombic field of the nucleus. It too is a two-body problem, but the reduced mass is almost the same as the electron mass. For the diatomic molecule the reduced mass is half the mass of either nucleus for homonuclear molecules but always on the order of nuclear masses, not the electronic mass. More importantly, the attractive force is not coulombic but rather harmonic; it becomes stronger as the distance between the nuclei gets larger. As a field theory professor of mine remarked to his class, all of quantum mechanics is the simple harmonic oscillator! He was not thinking of the diatomic molecule when he said that, but the diatomic molecule affords an excellent example of a system for which the simple harmonic oscillator is a good approximation.

When studying an electron in the hydrogen atom one can readily imagine the spherical coordinates used to locate the electron from the origin which is positioned at the proton (more properly at the center of mass but that is very close to where the proton is). For a diatomic molecule one should picture a dumbbell with the center of mass located midway between the nuclei for a homonuclear molecule, otherwise closer to the more massive nucleus. The dumbbell can now rotate about the center of mass, and only two angular parameters are needed to describe that. They are the same two as used for the angular description of the electron in hydrogen, and hence the rotational motion is a solved problem.

Imagining a diatomic molecule as a rigid dumbbell when rotating and as two masses connected to a spring when vibrating is simplistic on the one hand and remarkably good on the other. Your task is to keep in mind when and how these

R.L. Brooks, *The Fundamentals of Atomic and Molecular Physics*, Undergraduate Lecture Notes in Physics, DOI 10.1007/978-1-4614-6678-9_6,
© Springer Science+Business Media New York 2013

approximations break down and what can be done about that. One of the easier ways to picture the situation is to realize that the simple harmonic oscillator demands a potential curve that looks like a parabola while the real potential curve looks more like the Morse potential, to be treated subsequently. The real potential curve supports only a finite number of vibrational levels (often fewer than 20) while a parabola supports an infinite number. As one gets closer to the top of the finite potential, vibrational spacings are nothing like that predicted by the simple expressions derived for the harmonic oscillator even when corrected for anharmonic effects.

Perhaps one surprising feature that results from applying quantum mechanics to diatomic molecules is that the spin of the nuclei plays a role. This is a consequence of spin statistics, is a purely quantum effect, and is one that I have always found nonintuitive.

This chapter tries to present the core material that one requires when doing spectroscopy of diatomic molecules. Far more has been left out than included, but the student should be left in a position to read the classic monograph, *Spectra of Diatomic Molecules* by Gerhard Herzberg.[1]

6.1   Basic Considerations

In the previous chapter the solution of the electronic motion in a diatomic molecule was discussed, which represents the first part in the two-part Born–Oppenheimer approximation. The second part considers the motion of the nuclei, which is represented by (5.3) on page 117:

$$\left[\sum_i -\frac{\hbar^2}{2M_i}\nabla_i^2 + E(X_i)\right] v(X_i) = \mathcal{E}v(X_i) \tag{5.3}$$

Here X_i are the coordinates of the nuclei of mass M_i. $E(X_i)$ is the potential curve obtained in the previous chapter which will look like one of the two in Fig. 5.4.

The first step toward a solution of (5.3) is to separate out the center of mass solution for a system with two nuclei. On the off chance you have not seen this done before, it will be done here.

$$\text{Let} \quad \vec{R}_{CM} = \frac{M_1\vec{r}_1 + M_2\vec{r}_2}{M_1 + M_2} \tag{6.1}$$

$$\vec{R}_{rel} = \vec{r}_1 - \vec{r}_2 \tag{6.2}$$

---

[1] *Spectra of Diatomic Molecules*, 2nd ed. Gerhard Herzberg, Van Nostrand Reinhold, N.Y., 1950.

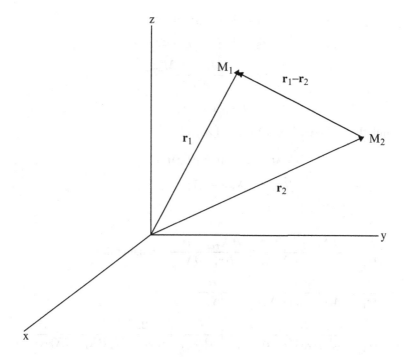

Fig. 6.1 Vectors relevant to the separation of the center of mass motion

$\vec{r}_1$ and $\vec{r}_2$ are the vectors from the origin to nucleus 1 and 2, respectively. Figure 6.1 presents the relevant vectors. Remember, all coordinates are now nuclear ones. The electron coordinates have all been integrated over to obtain the potential energy curve for nuclear motion.

Equation (5.3) may be written

$$\left[-\frac{\hbar^2}{2M_1}\nabla_1^2 - \frac{\hbar^2}{2M_2}\nabla_2^2 + E(|\vec{R}_{rel}|) \right] v(\vec{r}_1, \vec{r}_2) = \mathcal{E}v(\vec{r}_1, \vec{r}_2) \qquad (6.3)$$

It is easiest to work in Cartesian coordinates, and the idea is to reexpress the above second-order differential equation in center of mass and relative coordinates defined above. The following is a bit tedious but not difficult:

$$\nabla_1^2 = \frac{\partial^2}{\partial x_1^2} + \frac{\partial^2}{\partial y_1^2} + \frac{\partial^2}{\partial z_1^2}$$

$$\nabla_2^2 = \frac{\partial^2}{\partial x_2^2} + \frac{\partial^2}{\partial y_2^2} + \frac{\partial^2}{\partial z_2^2}$$

$$\vec{r}_1 = \hat{x}x_1 + \hat{y}y_1 + \hat{z}z_1$$

$$\vec{r}_2 = \hat{x}x_2 + \hat{y}y_2 + \hat{z}z_2$$

$$X_{CM} = \frac{M_1 x_1 + M_2 x_2}{M_1 + M_2}$$

$$X_{rel} = x_1 - x_2$$

and similarly for $Y_{CM}, Z_{CM}, Y_{rel},$ and $Z_{rel},$

$$\vec{R}_{CM} = \hat{x}X_{CM} + \hat{y}Y_{CM} + \hat{z}Z_{CM}$$

$$\vec{R}_{rel} = \hat{x}X_{rel} + \hat{y}Y_{rel} + \hat{z}Z_{rel}$$

Then

$$\frac{\partial}{\partial x_1} = \frac{\partial X_{CM}}{\partial x_1}\frac{\partial}{\partial X_{CM}} + \frac{\partial X_{rel}}{\partial x_1}\frac{\partial}{\partial X_{rel}} \quad \text{(chain rule)}$$

$$\frac{\partial}{\partial x_1} = \frac{M_1}{M_1 + M_2}\frac{\partial}{\partial X_{CM}} + \frac{\partial}{\partial X_{rel}}$$

$$\frac{\partial^2}{\partial x_1^2} = \frac{M_1^2}{(M_1 + M_2)^2}\frac{\partial^2}{\partial X_{CM}^2} + \frac{\partial^2}{\partial X_{rel}^2} + \frac{2M_1}{M_1 + M_2}\frac{\partial^2}{\partial X_{rel}\partial X_{CM}}$$

and similarly for $\frac{\partial^2}{\partial y_1^2}$ and $\frac{\partial^2}{\partial z_1^2}$. Also

$$\frac{\partial^2}{\partial x_2^2} = \frac{M_2^2}{(M_1 + M_2)^2}\frac{\partial^2}{\partial X_{CM}^2} + \frac{\partial^2}{\partial X_{rel}^2} - \frac{2M_2}{M_1 + M_2}\frac{\partial^2}{\partial X_{rel}\partial X_{CM}}$$

and similarly for $\frac{\partial^2}{\partial y_2^2}$ and $\frac{\partial^2}{\partial z_2^2}$.

Looking now at the x-components and remembering that precisely the same relations hold for y- and z-components, it is possible to write

$$\frac{\nabla_1^2}{M_1} \Rightarrow \frac{1}{M_1}\frac{\partial^2}{\partial x_1^2}; \qquad \frac{\nabla_2^2}{M_2} \Rightarrow \frac{1}{M_2}\frac{\partial^2}{\partial x_2^2}$$

Adding these gives

$$\frac{\nabla_1^2}{M_1} + \frac{\nabla_2^2}{M_2} \Rightarrow \frac{1}{M_1 + M_2}\frac{\partial^2}{\partial X_{CM}^2} + \frac{M_1 + M_2}{M_1 M_2}\frac{\partial^2}{\partial X_{rel}^2}$$

Let $M \equiv M_1 + M_2$ and $\mu \equiv \frac{M_1 M_2}{M_1 + M_2}$, and realizing that the as-yet-unsolved wave function can just as easily be considered a function of $\vec{R}_{rel}$ and $\vec{R}_{CM}$ yields

$$\left[-\frac{\hbar^2}{2M} \nabla^2_{CM} - \frac{\hbar^2}{2\mu} \nabla^2_{rel} + E(|\vec{R}_{rel}|) \right] v(\vec{R}_{rel}, \vec{R}_{CM}) = \mathcal{E}v(\vec{R}_{rel}, \vec{R}_{CM})$$

Assume separability, that is, that v may take the form

$$v(\vec{R}_{rel}, \vec{R}_{CM}) = v_{CM}(\vec{R}_{CM})v_{rel}(\vec{R}_{rel})$$

$$\text{and} \quad \mathcal{E} = \mathcal{E}'_{rel} + \mathcal{E}'_{CM}$$

The equation itself separates into the form

$$\frac{1}{v_{rel}} \left[-\frac{\hbar^2}{2\mu} \nabla^2_{rel} + E(R_{rel}) \right] v_{rel} - \mathcal{E}'_{rel} = \frac{1}{v_{CM}} \left[\frac{\hbar^2}{2M} \nabla^2_{CM} v_{CM} \right] + \mathcal{E}'_{CM}$$

Each side is clearly a function of different variables, so each may be set equal to a constant. Including that constant with the energy leads to the two equations

$$\left[-\frac{\hbar^2}{2\mu} \nabla^2_{rel} + E(R_{rel}) \right] v_{rel} = \mathcal{E}_{rel} v_{rel} \tag{6.4}$$

$$-\frac{\hbar^2}{2M} \nabla^2_{CM} v_{CM} = \mathcal{E}_{CM} v_{CM} \tag{6.5}$$

The second equation is just the Schrödinger equation of a free particle of mass M and is of no particular interest. The first equation is the Schrödinger equation for a "particle" of mass μ in the potential given by $E(R_{rel})$ and represents our looked for result.

The subscript rel can be dropped and (6.4) can be considered in some detail. Spherical coordinates may be used, r, θ, and ϕ, where $r \equiv R_{rel}$ is the distance between the nuclei and θ, ϕ are the usual spherical coordinates which determine the direction of the "dumbbell" molecule. The notation can be simplified somewhat by redefining

$$E(R_{rel}) \equiv V(r)$$

$$v_{rel}(r, \theta, \phi) = Z(r)Y_\ell^m(\theta, \phi)$$

This second equation is just a separation of coordinates which will look very similar to the hydrogen atom in that there is a single particle in a central potential. (The form of the potential is very different from hydrogen, so a very different radial solution can be expected. The angular solution, however, must be the same.) Then writing out the Laplacian in spherical coordinates (6.4) becomes

$$\left[-\frac{\hbar^2}{2\mu} \left(\frac{1}{r^2} \frac{\partial}{\partial r} \left[r^2 \frac{\partial}{\partial r} \right] - \frac{L^2}{\hbar^2 r^2} \right) + V(r) \right] Z(r) Y_\ell^m(\theta, \phi)$$

$$= EZ(r) Y_\ell^m(\theta, \phi)$$

Once again separating variables, calling the separation constant $\ell\,(\ell + 1)$ as was done for hydrogen, yields

$$\left[-\frac{\hbar^2}{2\mu} \frac{1}{r^2} \frac{d}{dr} \left(r^2 \frac{d}{dr} \right) + V(r) + \frac{\hbar^2 \ell\,(\ell + 1)}{2\mu r^2} \right] Z(r) = EZ(r) \qquad (6.6)$$

$$\text{and} \quad \mathbf{L}^2 Y_\ell^m((\theta, \phi)) = \ell\,(\ell + 1)\,\hbar^2 Y_\ell^m((\theta, \phi)) \qquad (6.7)$$

Equation (6.6) is the DE for the radial variable. The term $\frac{\hbar^2 \ell(\ell+1)}{2\mu r^2}$ comes from the angular part of the equation, so it is natural to associate this term with the rotational energy of the molecule. This is more clearly seen by noting that

$$\vec{r} = |\vec{r}_1 - \vec{r}_2| = [r_1^2 + r_2^2 - 2r_1 r_2 \cos(180°)]^{1/2} = r_1 + r_2$$

Since the origin is at the center of mass and $\vec{r}_1$ and $\vec{r}_2$ are oppositely directed, it follows that

$$M_1 r_1 = M_2 r_2 = \mu r$$

$$\text{So} \quad M_1 r_1^2 + M_2 r_2^2 = \frac{\mu^2 r^2}{M_1} + \frac{\mu^2 r^2}{M_2} = \mu r^2$$

So $\mu r^2 = M_1 r_1^2 + M_2 r_2^2 = I$, which is the moment of inertia of the dumbbell molecule. Then

$$E_{rot} = \frac{\hbar^2 \ell\,(\ell + 1)}{2I} \qquad (6.8)$$

It is tempting to treat I as a constant which would certainly simplify things. In fact, this is done as a first approximation to a solution. Remember that r changes but little about some equilibrium value. Such small changes in r produce changes in E_{rot} much smaller than changes in $V(r)$. To a first approximation if E_{rot} is dropped from the Hamiltonian, the energy due to vibrations could be found, and then the rotational energy could be added back in. This shall be done as a first attempt at a solution. Later it will be seen that inclusion of this term does make second-order but easily measurable changes to molecular energy levels.

With those comments in mind, return to (6.6) and write this for $E_{rot} = 0$. (This is no approximation when $\ell = 0$.) What this produces is the differential equation for vibrational motion which will next be considered. The rotational energy will be added back in afterward which is equivalent to treating it as a constant:

$$\left[-\frac{\hbar^2}{2\mu} \frac{1}{r^2} \frac{d}{dr} \left(r^2 \frac{d}{dr} \right) + V(r) \right] Z(r) = EZ(r) \qquad (6.9)$$

Consider solutions close to the equilibrium separation r_0, where $V(r)$ achieves a minimum. Taking the zero of potential to be at infinite nuclear separation, then $V(r_0) \equiv -D_e$, which is the equilibrium dissociation energy. Next expand $V(r)$ in a Taylor series about r_0:

$$V(r) \approx V(r_0) + \left.\frac{\mathrm{d}}{\mathrm{d}r}\right|_{r=r_0}(r-r_0) + \left.\frac{\mathrm{d}^2 V}{\mathrm{d}r^2}\right|_{r=r_0}\frac{(r-r_0)^2}{2!}\cdots$$

But $V(r)$ has a minimum at $r=r_0$ so $\left.\frac{\mathrm{d}V}{\mathrm{d}r}\right|_{r=r_0} = 0$. Call $\left.\frac{\mathrm{d}^2 V}{\mathrm{d}r^2}\right|_{r=r_0} \equiv k$. Then

$$V(r) \approx -D_e + \frac{k}{2}(r-r_0)^2 = -D_e + \frac{k}{2}x^2 \tag{6.10}$$

and $x \equiv r - r_0$.

What is about to be done is the transformation of the original differential equation, written for a function $Z(r)$, into one written for a function $H(x)$, with x defined above. The reason is simply to try to put the differential equation into a recognizable form with a known solution. In this case, the solution shall involve the Hermite polynomials.

The differential operation in Eq. (6.9) may now be written

$$\frac{1}{r^2}\frac{\mathrm{d}}{\mathrm{d}r}\left(r^2\frac{\mathrm{d}Z}{\mathrm{d}r}\right) = \frac{\mathrm{d}^2 Z}{\mathrm{d}r^2} + \frac{2}{r}\frac{\mathrm{d}Z}{\mathrm{d}r} = \frac{\mathrm{d}^2 Z}{\mathrm{d}x^2} + \frac{2}{x+r_0}\frac{\mathrm{d}Z}{\mathrm{d}x}$$

As was done for hydrogen, let $H(x) \equiv (x+r_0)Z(x+r_0)$. Then

$$\frac{\mathrm{d}H}{\mathrm{d}x} = (x+r_0)\frac{\mathrm{d}Z}{\mathrm{d}x} + Z$$

$$\frac{\mathrm{d}^2 H}{\mathrm{d}x^2} = (x+r_0)\frac{\mathrm{d}^2 Z}{\mathrm{d}x^2} + 2\frac{\mathrm{d}Z}{\mathrm{d}x}$$

and

$$\frac{1}{x+r_0}\frac{\mathrm{d}^2 H}{\mathrm{d}x^2} = \frac{\mathrm{d}^2 Z}{\mathrm{d}x^2} + \frac{2}{x+r_0}\frac{\mathrm{d}Z}{\mathrm{d}x}$$

This then leads to

$$\frac{\mathrm{d}^2 H}{\mathrm{d}x^2} + \frac{2\mu}{\hbar^2}\left(E' - \tfrac{1}{2}kx^2\right)H = 0 \tag{6.11}$$

where $E' \equiv E + D_e$.

Equation (6.11) is the Schrödinger equation of a one-dimensional simple harmonic oscillator in the displacement x from equilibrium. The functions $H_n(x)$ are Hermite polynomials times an exponential, and the solution for the eigenvalues is

$$E'_n = \hbar\sqrt{\frac{k}{\mu}}(n+\tfrac{1}{2}) = \hbar\omega(n+\tfrac{1}{2})$$

with $\omega^2 \equiv k/\mu$. So the solution for the vibrational energy levels is given by

$$E_{vib} = -D_e + \hbar\omega(n + {}^1\!/_2) \tag{6.12}$$

Including the rotational energy levels, the full solution may be written

$$E_{tot} = -D_e + \hbar\omega(n + {}^1\!/_2) + \frac{\hbar^2 \ell(\ell + 1)}{2I} \tag{6.13}$$

A review of the levels of approximation which were used to obtain this result would be useful. The Born–Oppenheimer separation is fundamental, is often quite good, and gives $V(r)$ as a solution to the electronic problem. Next a centrifugal force contribution to the potential was identified as the rotational energy. This is not all that bad an approximation. Finally the potential curve $V(r)$ was approximated by a parabola which, while not bad for very small n, is expected to be rather poor as n becomes large. Doing better than making a parabolic approximation to the potential will be the next order of business.

6.2   The Anharmonic Oscillator and Nonrigid Rotator

There exists an analytic potential function with a single adjustable parameter (besides the well depth) which can reasonably approximate most ground-state potential curves. In addition it has the property that the Schrödinger equation can be exactly solved for this potential. It is called the Morse potential and is given by

$$V(r) = D_e \left[e^{-2a(r-r_0)} - 2e^{-a(r-r_0)} \right] + A \tag{6.14}$$

Here a is an adjustable shape parameter and D_e is the depth of the well at the equilibrium separation below the asymptotic value of A.

Figure 6.2 shows a Morse potential for H_2 with $r_0 = 1.4011a_0$, $a = 0.9790a_0^{-1}$, $D_e = 0.1745$ a.u., and $A = -1.0$ a.u. It really is quite good for rough work, but one should take notice that any Morse potential does poorly for small values of r. At $r = 0$ the potential is finite rather than infinite as Coulomb repulsion would demand.

The Morse potential yields the following values for the vibrational energy in place of (6.12) of the previous section:

$$E_{vib} = -D_e + \hbar\omega(n + {}^1\!/_2) - \frac{\hbar^2\omega^2}{4D_e}(n + {}^1\!/_2)^2 \tag{6.15}$$

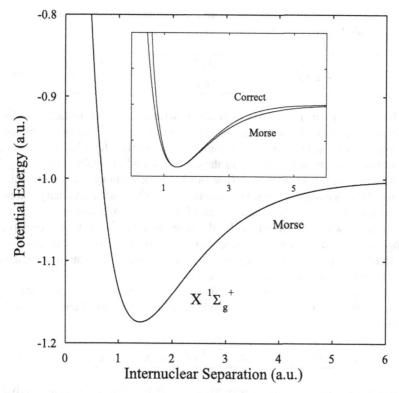

Fig. 6.2 Morse potential for H_2 with the "correct" potential of the previous chapter shown in inset using the same scale for comparison

It is well known in quantum mechanics that such a quadratic arises from the treatment of the anharmonic oscillator. In such a treatment, $V(r)$ is expressed as a power series in $(r - r_0)$, and terms higher than $(r - r_0)^2$ are treated as perturbations of the harmonic oscillator. Such a treatment results in terms higher than $(n + \frac{1}{2})^2$. For most diatomic molecules, (6.15) accounts for the energy separation reasonably well up to modest values of n. The Morse potential itself has the added advantage of yielding a reasonable potential curve for fairly large values of $(r - r_0)$.

How good is Eq. (6.15) for hydrogen? At $v = 0$ the formula yields a value only 0.3 % above that of a good quantum mechanical calculation. The error increases smoothly with v, becoming 11 % at $v = 5$ and 40 % at $v = 10$.

Problem 6.1

Express the vibrational frequency ω in terms of a, D_e, and μ of the Morse potential.

Problem 6.2

How many vibrational levels does Eq. (6.15) predict are bound for molecular hydrogen? Compare to the correct value of 14.

The energy levels of a diatomic molecule with anharmonic vibration but rigid rotation could then be expressed as

$$E_{tot} = -D_e + \hbar\omega(n + \tfrac{1}{2}) - \frac{\hbar^2\omega^2}{4D_e}(n + \tfrac{1}{2})^2 + \frac{\hbar^2 J(J + 1)}{2I} \qquad (6.16)$$

The angular momentum for a molecule is written as J, but in fact J includes more rotational contributions than just the mechanical tumbling of the molecule taken up previously. The mechanical tumbling is properly referred to as N, and this is coupled to the angular momenta of the electron cloud to obtain the total angular momentum always referred to as J. How this is done gets tricky as one distinguishes different Hund's coupling cases. Our discussion is sufficiently elementary that none of these complications will be taken up and for $^1\Sigma$ states do not exist.

The classic work on diatomic molecules is by Herzberg,[2] and spectroscopists use strange notation. It is worth some space here to rewrite some of our expressions the way Herzberg writes them so that you may take advantage of extensive compilation of the known constants for diatomic molecules.

First, express all energies in cm^{-1}. The symbol ω means $1/\lambda$. The zero of energy is taken at the minimum of the potential curve. The vibrational energies are then written

$$G(v) = \omega_e(v + \tfrac{1}{2}) - \omega_e x_e(v + \tfrac{1}{2})^2 + \omega_e y_e(v + \tfrac{1}{2})^3 + \cdots \qquad (6.17)$$

Here v takes the place of n previously. Clearly this expression is *not* derivable from a Morse curve but rather from a power expansion of $V(r)$.

Using this same notation the rotational energies are written

$$F_v(J) = B_v J(J + 1) - D_v J^2(J + 1)^2 + \cdots \qquad (6.18)$$

with

$$B_v \equiv B_e - \alpha_e(v + \tfrac{1}{2}) \qquad (6.19)$$

$$D_v \equiv D_e - \beta_e(v + \tfrac{1}{2}) \qquad (6.20)$$

and

$$B_e \equiv \frac{\hbar}{4\pi cI} \qquad D_e = \frac{4B_e^3}{\omega_e^2}$$

---

[2]*Spectra of Diatomic Molecules*, 2nd ed. Gerhard Herzberg, Van Nostrand Reinhold, N.Y. 1950.

B_e, α_e, and β_e would be tabulated. Clearly the first term in (6.18) is the same as our E_{rot} written in wave number energy units, except that it allows for a slight change in the moment of inertia as the molecule rises to high vibrational levels. D_v is a *very* small correction and can usually be ignored.

Example: The N_2 *Molecule*

For the ground electronic state of the nitrogen molecule, N_2, the following numbers can be found in Herzberg[3]:

$$\omega_e = 2359.6 \text{ cm}^{-1} \qquad B_e = 2.01 \text{ cm}^{-1}$$
$$\omega_e x_e = 14.46 \text{ cm}^{-1} \qquad \alpha_e = 0.019 \text{ cm}^{-1}$$
$$\omega_e y_e = 0.0075 \text{ cm}^{-1} \qquad D_e = 6 \cdot 10^{-6} \text{ cm}^{-1}$$
$$\beta_e = 10^{-9} \text{ cm}^{-1}$$

One sees immediately that the dropping of vibrational contributions from terms higher than quadratic in $(n + 1/2)$ is justified for 1% accuracy. Also, ignoring the rotational contribution of D_v is justified.

The moment of inertia for N_2 is given by

$$I = \frac{\hbar}{4\pi c B_e} = 29.9 \text{ amu } r_B^2$$

The force constant k_e is given by

$$k_e = 5.88 \cdot 10^{-2} \mu \omega_e^2 \frac{\text{dyne}}{\text{cm}} = 2.30 \cdot 10^{+6} \text{ dy/cm} \quad \text{for } N_2$$
$$= 157 \text{ lbs/ft} \quad (!)$$

6.3 Transitions and Selection Rules

When considering transitions between states of molecules one must carefully distinguish between electronic transitions in which one electron "jumps" to a different molecular orbital, thereby jumping between potential curves, and transitions within the same potential curve in which the electron does not take part.

---

[3]More recent numbers are compiled in *Constants of Diatomic Molecules*, K.P. Huber and G. Herzberg, Van Nostrand Reinhold, N.Y. 1979.

The former transitions are completely analogous to the atomic case, in which an electric dipole is induced by the electromagnetic field of a photon. The selection rules are likewise analogous and arise precisely the same way they do for atoms. The details are beyond the scope of the present treatment. Calling $\Lambda = M_L$, there is the rule for electronic transitions

$$\Delta\Lambda = 0, \pm 1.$$

The additional symmetries, $\pm$ and g, u, behave according to

$$g \leftrightarrow u \qquad g \nleftrightarrow g \qquad u \nleftrightarrow u$$
$$\Sigma^+ \leftrightarrow \Sigma^+ \qquad \Sigma^- \leftrightarrow \Sigma^- \qquad \Sigma^+ \nleftrightarrow \Sigma^-$$

The vibrational rule is $\Delta v =$ anything (for different electronic states). The rotational rule is $\Delta J = 0, \pm 1$ but not $0 \to 0$. For transitions in the infrared, in which the electron does not jump but stays in the ground-state potential, it is not obvious that *any* transitions will occur.

When the two nuclei have different charges, it seems reasonable to suppose that the electrons will localize in such a way that the molecule will possess a permanent dipole moment, $\vec{P}$. Such a dipole moment will represent an additional term in the Hamiltonian which so far has been ignored. The radiation field $\vec{A}$ will interact with this dipole moment in the same way that it interacts with the electronic moment so that transitions will be proportional to

$$\langle \psi_i | \vec{P} | \psi_f \rangle$$

Now assuming that $\vec{P}$ is proportional to the internuclear separation and directed along that axis and that $x = r - r_e$, as before, the dipole moment may be expanded as

$$P(r) = P(x + r_e) \equiv P(r_e) + x \left. \frac{dP}{dr} \right|_{r=r_e} \equiv \mathbf{P}_0 + x\mathbf{P}_1$$

For harmonic oscillator wave functions this leads to

$$\langle \psi_i | \mathbf{P} | \psi_f \rangle = \langle H_{v'} | \mathbf{P}_0 + x\mathbf{P}_1 | H_{v''} \rangle$$

where $H_{v'}$ is the Hermite polynomial of the lower state with eigenvalues v' and $H_{v''}$ is for the upper state. Note that the operator $x\mathbf{P}_1$ is proportional to the rate of change of the dipole moment with internuclear separation. Using the recursion relation

$$xH_v(x) = {}^1\!/_2 H_{v+1}(x) + vH_{v-1}(x)$$

the selection rule follows immediately

$$\Delta v = \pm 1.$$

for vibrational states belonging to the same electronic state. The rotational selection rule is

$$\Delta J = 0, \pm 1 \quad \text{but not} \quad 0 \to 0.$$

Of course the selection rule $\Delta v = \pm 1$ is only as good as the harmonic oscillator approximation. When anharmonic terms are included, the rule can be violated but usually with a significantly reduced intensity.

It follows from this discussion that homonuclear molecules have *no dipole transitions*. Very weak transitions can occur if the quadrupole of neighboring molecules can induce a dipole on a given molecule, which then undergoes an allowed vibrational or rotational transition. Such transitions become easily observable in dense gases and in certain liquids and solids.

Of course selection rules tell you which transitions are forbidden. But among allowed transitions there is an extreme variability of intensities. Consider a transition between two different electronic potential curves, one that occurs for the same reasons that excited atoms make transitions. The electronic dipole moment, μ^e, is the relevant transition operator, and calculating the transition moment is not easy. For the point being made now, it is not important. Realize that the operator itself is a function only of electronic coordinates. The transition moment between levels i and j is proportional to

$$\begin{aligned} A_{ij} &\propto [\langle \psi_i | \mu^e | \psi_j \rangle]^2 = [\langle u_i v'' | \mu^e | u_j v' \rangle]^2 \\ &= [\langle u_i | \mu^e | u_j \rangle \langle v'' | v' \rangle]^2 \end{aligned} \qquad (6.21)$$

where the details are unimportant so don't be concerned about calculating these terms. What is relevant is to notice that the total transition moment is a function of both the electronic and nuclear (vibrational) coordinates. The last term, $|\langle v'' | v' \rangle|^2$, which is a function of only the nuclear coordinates, is an overlap integral of the vibrational wave functions of the upper state and lower state involved in the transition. It is called the Franck–Condon factor and can be large whenever the two potential curves are of similar shape and line up, one on top of the other and $\Delta v = 0$. Guessing whether or not this factor is large by inspection of the potential curves is often done but probably shouldn't be as detailed calculations are really needed to make an appropriate assessment.

The keen-eyed student will have noticed that the separation that gave rise to this factor for the vibrational wave functions would have done something similar for the rotational wave functions for precisely the same reason. This factor has not been included above (but could have been) and is called the Hönl–London factor. More properly, the Hönl–London factor is more complicated than just a simple overlap integral. The issue of laboratory-fixed versus molecule-fixed reference frames is relevant, and such details are beyond this elementary treatment. This factor is not so important for survey work because typically either an entire rotational branch shows in a spectrum or it doesn't and it is impossible to estimate this factor on inspection of the potential curves. It would be included when doing detailed calculations.

6.4 Thermal Distribution of Quantum States

Consider the lowest few vibrational and rotational states of a diatomic molecule. The vibrational energy levels may be written as

$$G(v) = \omega_e(v + \tfrac{1}{2}) - \omega_e x_e(v + \tfrac{1}{2})^2 \tag{6.22}$$

$$G(v = 0) = \frac{\omega_e}{2} - \frac{\omega_e x_e}{4} \tag{6.23}$$

$G(v = 0)$ is the energy of the ground vibrational state with respect to the minimum of the potential well. Let

$$G_0(v) \equiv G(v) - G(v = 0) = \omega_e v - \omega_e x_e(v^2 + v) \tag{6.24}$$

For a Maxwell–Boltzmann distribution the number of molecules with energy $G_0(v)$ will be distributed according to $e^{-E/kT}$. Or

$$\frac{N_v}{N_0} = e^{-G_0(v)hc/kT} = e^{-G_0(v)/0.6952T} \tag{6.25}$$

the latter form being correct when G_0 is in cm^{-1} and T is in Kelvin. The total number of molecules must be

$$N = \sum_{v=0}^{\infty} N_v = N_0 \sum_{v=0}^{\infty} e^{-G_0(v)hc/kT} \equiv N_0 Q_v \tag{6.26}$$

Q_v is called the state sum or partition function. The fraction of molecules having energy $G_0(v)$ must then be given by

$$\boxed{\frac{N_v}{N} = \frac{e^{-G_0(v)hc/kT}}{Q_v}} \tag{6.27}$$

This is simply the fraction of molecules having a vibrational quantum number of v. For example, looking at N_2, using the values given on page 149, N_1/N at 300 K is given by 1.17×10^{-5} while at 1,000 K $N_1/N = 3.21 \times 10^{-2}$. So at room temperature 99.99 % of the N_2 molecules are in the ground vibrational level. For this example, the result does not change much if one looks at N_1/N_0 which is easier to calculate. For heavier molecules that is not the case.

Since $w_e \propto 1/\sqrt{\mu}$ for heavy diatomic molecules, the fraction N_1/N (as well as N_1/N_0) will be larger. Indeed, for I_2 at 300 K, $N_1/N = 0.23$, and a very large fraction (36 %) of molecules are not in the ground state.

Problem 6.3

Using values for I_2, $w_e = 214.57$ and $w_e x_e = 0.6127$ verify the above statement.

An analogous calculation may be performed for the distribution of rotational levels. There is, however, one big difference which is that each rotational level has a $(2J+1)$ degeneracy. Writing

$$F_v(J) = B_v J(J+1) \quad \text{(neglecting } D_v) \tag{6.28}$$

the ratio of the population N_J/N_0 (for a given fixed v) is

$$\frac{N_J}{N_0} = (2J+1)e^{-F_v(J)hc/kT} \tag{6.29}$$

Defining the rotational partition function to be

$$Q_r \equiv \sum_{J=0}^{\infty} (2J+1)e^{-B_v J(J+1)hc/kT} \tag{6.30}$$

the fraction of total molecules in the rotational state J is given as before by

$$\boxed{\frac{N_J}{N} = \frac{(2J+1)e^{-B_v J(J+1)hc/kT}}{Q_r}} \tag{6.31}$$

The behavior of this distribution as a function of J is fundamentally different from the simple exponential behavior of the analogous distribution for vibration. This one for rotations has a maximum. Its behavior as a function of J is dictated by the numerator of (6.31). The curve for the ground vibrational state of N_2 looks like the following:

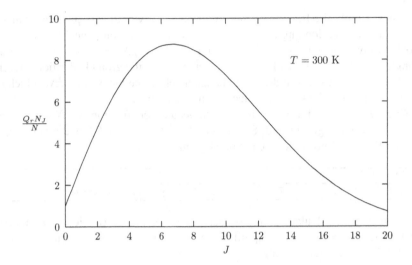

The value of the maximum is readily seen to be

$$J_{max} = \sqrt{\frac{kT}{2B_v hc}} - \frac{1}{2} = 0.5896 cm^{-1/2}/K^{1/2}\sqrt{\frac{T}{B_v}} - \frac{1}{2}$$

Problem 6.4

Derive this expression for J_{max}.

It can be seen that even for very low temperatures (~ 4 K), N_2 will have a number of excited rotational states. However, for molecules like H_2 ($B_0 = 60$ cm^{-1}) or D_2 ($B_0 = 30$ cm^{-1}), virtually all of the molecules are in the ground rotational state at 4 K. In the next section it will be necessary to modify this conclusion when nuclear spin is considered.

For sufficiently large T or small B_v, the sum in (6.30) can be replaced by an integral yielding the approximation

$$Q_r \approx \int_0^\infty (2J + 1)e^{-B_v J(J+1)hc/kT} \, dJ \tag{6.32}$$

$$\text{Let} \quad R \equiv B_v hc/kT$$

$$x \equiv J(J + 1)$$

$$dx \equiv (2J + 1) \, dJ$$

$$\text{then} \quad Q_r = \int_0^\infty e^{-Rx} \, dx = 1/R$$

So

$$Q_r = \frac{kT}{B_v hc} \quad \text{(High temp.)} \tag{6.33}$$

Equation (6.31) then becomes

$$\boxed{\frac{N_J}{N} \approx \frac{B_v hc}{kT}(2J+1)e^{-B_v J(J+1)hc/kT}} \tag{6.34}$$

for high temperature or small B_v. For $T \geq 300$ K and $B \leq 10$ cm^{-1}, (6.33) will be better than 2 % as an approximation to (6.30).

6.5   Effects of Nuclear Spin for Homonuclear Molecules

Recall that in the Born–Oppenheimer approximation, the total wave function looks like

$$\psi = \psi_e \psi_v \psi_r \chi \tag{6.35}$$

where e means electronic, v means vibrational, and r means the rotational component of the total wave function. χ is the spinor which will be treated subsequently.

The question that needs to be answered is the following: what happens to the wave function if two identical nuclei are interchanged? The answer depends on whether the nuclei are fermions or bosons. This is determined by the total nuclear spin I. If it is half-integer, the nucleus is a fermion; if it is integer, the nucleus is a boson. For fermions, the wave function must change sign while for bosons it must not change sign.

The task at hand is to determine what happens to the sign of the above wave function under interchange of nuclei. An exchange of nuclei can result from a double operation. First invert the entire wave function, nuclei, and electrons; then again invert just the electrons. First, consider just the coordinate-dependent wave functions, ψ_e, ψ_v, and ψ_r; the spinor will be treated afterward. χ is not a function of the usual spatial coordinates.

The *total* coordinate wave function is said to be positive or negative depending on whether it changes sign under inversion of all coordinates. ψ_v never changes sign, so there is no worry about it. ψ_r changes sign under inversion depending on whether J is even or odd. For ψ_e, it depends on whether Λ is zero (a Σ state) or nonzero. If nonzero, there is a double degeneracy, such that one state changes sign and the other does not; hence this symmetry operation *imposes no* constraints on the wave function.

For Σ states, a total inversion effectively changes the direction of the internuclear axis and hence changes sign for Σ^- states while Σ^+ states remain unaffected. So consider the consequences of the first operation—total inversion of the nuclear and electronic coordinates:

$$\Sigma^+ \qquad\qquad\qquad \Sigma^-$$

J			J		
3	____	$-$	3	____	$+$
2	____	$+$	2	____	$-$
1	____	$-$	1	____	$+$
0	____	$+$	0	____	$-$

The $+$ or $-$ to the right of every level represents the product of the evenness or oddness of the J value with the superscript of the Σ. If the state were Π or Δ, for example, none of this discussion is relevant because the double degeneracy contains states of both symmetries so anything is possible. Now this isn't a true statement always because high-resolution work can often resolve what is called lambda doubling, but for the purpose of this discussion it will be accepted.

The second operation now is to invert just the electronic wave functions. Such an inversion will change the sign of Σ_u wave functions but leave Σ_g ones unaffected. Call the total coordinate wave functions symmetric (s) or antisymmetric (a) depending on whether it changes sign after *both* these operations are performed, these operations being equivalent to an exchange of nuclei:

$$\Sigma_g^+ \qquad\qquad\qquad \Sigma_u^+$$

J					
3	____	$-\ a$		____	$-\ s$
2	____	$+\ s$		____	$+\ a$
1	____	$-\ a$		____	$-\ s$
0	____	$+\ s$		____	$+\ a$
	(a)			(b)	

$$\Sigma_g^- \qquad\qquad\qquad \Sigma_u^-$$

J					
3	____	$+\ s$		____	$+\ a$
2	____	$-\ a$		____	$-\ s$
1	____	$+\ s$		____	$+\ a$
0	____	$-\ a$		____	$-\ s$
	(c)			(d)	

The diagrams above are twice as numerous as previously because each of the Σ states may be "g" or "u" in addition to "$+$" or "$-$". "a" or "s" to the right of each level is the product of the $+$ or $-$ at the right of each level with the evenness or oddness of "g" or "u". It represents the total coordinate symmetry, under exchange of nuclei, of the state under consideration.

In order to tell whether the total wave function is symmetric or antisymmetric to an exchange of nuclei, one needs to consider the spin wave functions and multiply their symmetry by that of the coordinate wave function.

If the two nuclei have spins of I_1 and I_2, then the total spin (call it T) is given by

$$T = I_1 + I_2 \tag{6.36}$$

where the addition is that of two angular momenta. For homonuclear molecules, the only ones that need to be considered here, $I_1 = I_2$, giving

$$T = 2I, 2I - 1, 2I - 2, \ldots, 0$$

Consider the case of H_2 in which each nucleus has a spin of $^1/_2$. The total nuclear spin of each molecule may be 0 or 1. The total nuclear spin eigenfunctions must be singlet and triplet and look, by now familiarly, like

Triplet	Singlet
$\alpha(1)\alpha(2)$	
$\alpha(1)\beta(2) + \alpha(2)\beta(1)$	$\alpha(1)\beta(2) - \alpha(2)\beta(1)$
$\beta(1)\beta(2)$	

α and β are the spin-up and spin-down spinors for spin $^1/_2$ particles. Consider the ground electronic state of H_2 which is Σ_g^+, so table (a) is relevant. Since the nuclei are fermions, the total wave function must be antisymmetric under exchange of nuclei. Since the coordinate functions are symmetric for $J = 0, 2, 4$, these must be paired with the antisymmetric singlet states while $J = 1, 3, 5$, etc., must be paired with the symmetric triplet states. The triplet nuclear spin states have a statistical weight of 3 ($M_T = -1, 0, +1$), while singlets have a statistical weight of 1. When the rotational spectrum is analyzed, the lines show an alternation of intensities in the ratio 3:1.

The states of a given symmetry with the greater statistical weight are called *ortho* while those with the lesser weight are called *para*. These species are noninteracting in the absence of external fields. For H_2, the rotational levels may be labeled:

$$H_2 \quad \Sigma_g^+$$

J		
3	_____	$- a$ *ortho*
2	_____	$+ s$ *para*
1	_____	$- a$ *ortho*
0	_____	$+ s$ *para*

Let's review the notation. The numbers on the left are the J values, the rotational quantum number. The $+, -$ on the right are the symmetry of the wave function to an inversion of all coordinates and are given by $(-1)^J$ multiplied by the $+, -$ superscript on $\Sigma^\pm$. The s and a on the right are the total symmetry of the coordinate wave function obtained by multiplying the $+, -$ on the right by $+1$ for gerade and -1 for ungerade. Finally *para* and *ortho* are determined by considering whether the nuclei are fermions or bosons. H_2 is composed of fermions, so the total wave function must be antisymmetric. Hence the antisymmetric (*para*) spin functions are matched to the symmetric coordinate functions, and the symmetric (*ortho*) spin functions are matched to the antisymmetric coordinate functions.

For nuclear spins greater than $^1/_2$, it is not so easy to know whether the total spin function is symmetric or antisymmetric. The following will help:

Nuclear spin symmetric if: $T = 2I, 2I - 2, 2I - 4, \ldots$	
Statistical weight is: $(2I + 1)(I + 1)$	(6.37)
Always ORTHO	
Nuclear spin antisymmetric if: $T = 2I - 1, 2I - 3, 2I - 5, \ldots$	
Statistical weight is: $(2I + 1)(I)$	(6.38)
Always PARA	

This discussion has been about a purely quantum effect which dictates whether or not a given rotational level of a homonuclear diatomic molecule can be populated at all and if so, with what statistical weight. These weights then can be readily verified by measurements of the relative intensities of rotationally resolved spectral lines. The more one reflects on the fact that a diatomic molecule, which tumbles because it is shaped like a dumbbell, obeys a quantum rule for tumbling dictated by the spin statistics of its constituent nuclei, the more one is impressed by the extraordinary success of quantum mechanics to describe the physical world. At least I am!

When solving problems from this section one needs to proceed in reverse order to how the material was presented. First consider whether the nuclei are fermions or bosons which dictates the overall symmetry of the total wave function. Then you need to obtain that symmetry by forming the product of the symmetry of the *ortho* or *para* states with the symmetry of the coordinate state of a given J value. The example below brings this altogether.

Example

Consider a $^{10}B_2$ molecule with $I = 3$ and a ground electronic state of Σ_g^-. The symmetric nuclear states have $T = 6, 4, 2$, and 0 while the antisymmetric ones are $T = 5, 3$, and 1. The statistical weight of the symmetric ones is 28. (One does not need formula (6.37) above. Just count the M_T values. There are $(2 \cdot 6 + 1) + (2 \cdot 4 + 1) + (2 \cdot 2 + 1) + (2 \cdot 0 + 1)$.) The statistical weight of the antisymmetric ones is 21. The nuclei are bosons, so

$$\text{B}_2 \qquad \Sigma_g^-$$
$$J$$

3	_____ + s ortho
2	_____ − a para
1	_____ + s ortho
0	_____ − a para

At very low temperatures, when all of the molecules sink to the lowest allowed states, $3/7$ will be in the $J = 0$ level, while $4/7$ will be in the $J = 1$.

Problem 6.5

Try N_2 ($I = 1$).

6.6 Labeling of Rotational Bands

Transition designators between levels in diatomic molecules use a special shorthand notation for the rotational aspect of the transition. This notation is used whether the transition occurs between two different electronic potential curves or simply between two different vibrational levels within a single electronic potential. Recall that the selection rule for J for a (dipole) transition is $\Delta J = 0, \pm1$. Transitions with ΔJ greater than ±1 can occur through quadrupolar and higher multipole moments, but such transitions become weaker as the multipole becomes higher. The transitions are labeled alphabetically starting from Q when $\Delta J = 0$. Here ΔJ means J of the upper level minus J of the lower level. Higher letters in alphabetical order are used for positive ΔJ and lower ones for negative ΔJ. That is,

$$Q \text{ when } \Delta J = 0$$

P when $\Delta J = -1$	R when $\Delta J = +1$
O when $\Delta J = -2$	S when $\Delta J = +2$
N etc.	T etc.

The J value of the level lower in energy is placed in parentheses, following the letter. Note that the rule states "lower level" which can be either the initial level or the final level depending on whether the transition occurs in absorption or emission.

So $R(1)$ is a transition from

$J = 1$	to $J = 2$	in absorption
$J = 2$	to $J = 1$	in emission

Look at the figure for $^4\text{He}_2$. Because this transition occurs in the visible part of the spectrum, one can assume that it is an electronic transition. Molecular helium is an excimer, which means that the lowest electronic potential curve, the X $^1\Sigma_g^+$, is repulsive; it has no significant potential well. (Determining the size of the very slight potential depression is an ongoing endeavor.) But once an electron is excited, there exist a series of potential wells with increasingly larger energy until one reaches the lowest potential for the He_2^+ molecular ion which is stable. Transitions can occur between these excited potential curves for molecular helium just as they can for any other diatomic molecule. Of course the main difference is that a spontaneous transition to the ground-state potential causes the excimer to fly apart. A discharge in helium gas will result in both molecular and atomic lines with higher gas pressure favoring the formation of excimer transitions. The spectra being presented (because they are extremely clean examples of the effects of nuclear spin on rotational transitions) were actually acquired from helium gas at 4.2 K being excited by a proton beam, but that need not be of concern here.[4]

The strong feature just above 15,600 cm^{-1} is the unresolved Q-branch of the $^3\Sigma_u^+ \rightarrow {}^3\Pi_g$ ($v = 0$ to $v = 0$) transition which has an R-branch to higher frequency and a P-branch to lower frequency. First take note that every second line is missing. This is a consequence of applying the rules for nuclear spin statistics to each of the potentials in turn. Only the upper, $^3\Sigma_u^+$, is restrictive, and because the nuclei are bosons, the total wave function must be symmetric. The Σ_u^+ wave functions are symmetric for odd J values, so only those can exist in the upper potential. When a transition is made to the lower potential, J changes by zero (Q-branch) or ± 1 (P- and R-branches), and every J value is possible in the lower potential as there are no symmetry restrictions because of the double degeneracy. But odd J values in the upper level convert to even J values in the lower level for both R- and P-branch transitions. Since such transitions are labeled by the lower J value one sees only even values on the figure.

Problem 6.6

If the transition were $^3\Sigma_u^+ \rightarrow {}^3\Sigma_g^+$, show that the Q-branch is missing.

If one looks at the same electronic transition in the isotopic molecule, $^3He_2$ quite a bit is different. First the nuclei are fermions and so the entire wave function must be antisymmetric. But the total nuclear spin can be 1 or 0 as the individual nuclei have spin $^1/_2$. The molecules with nuclear spin 1 are *ortho*, have a statistical weight of 3, and have a symmetric wave function. Those with nuclear spin of 0 are *para*, have a statistical weight of 1, and have an antisymmetric wave function. For the same reasons as previously, only the upper, $^3\Sigma_u^+$, is restrictive, but now even J values must be paired with the symmetric *ortho* spin states while odd J values are paired with the *para* states in order for the entire wave function to be antisymmetric. The lower electronic level is nonrestrictive because it is Π, but after a transition, the molecules that started with an even J value find themselves with an odd value following either a P or R transition. These are the ones that came from the levels

---

[4]R.L. Brooks and J.L. Hunt, J.Chem.Phys. **88**, 7267 (1988).

with the higher statistical weight, so the stronger lines have labels in the figure with odd J values. The ratio of intensities is 3 to 1 which is the ratio of statistical weights dictated by the nuclear spin statistics.

6.7 Transitions Revisited

The previous two figures present an opportunity to point out a rather fundamental aspect of molecular spectroscopy. Molecular constants, the ones tabulated in books or online sources, almost always are derived from spectra not from ab initio calculations. Extraction of these constants from detailed line-lists is done using least-squares fitting on any modest computer. It is useful for you to realize that perfectly reasonable estimates for some of these constants can be obtained by taking judicious differences of wave number values and using elementary logic.

For example, if one wanted to obtain the rotational constant for the lower $^3\Pi$ and vibrational quantum number 0 (because the transitions in Figs. 6.3 and 6.4 are for $v = 0$ to $v = 0$ in each potential curve), one could subtract two transition values that originate on the same upper level but have different lower levels. For example, the $R2$ and $P4$ transitions both have $J = 3$ in their upper originating energy level. The difference between the two quoted transition values would then be the energy

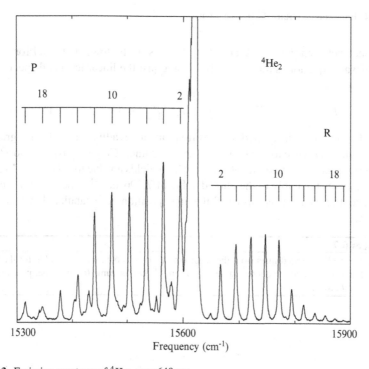

Fig. 6.3 Emission spectrum of $^4He_2$ near 640 nm

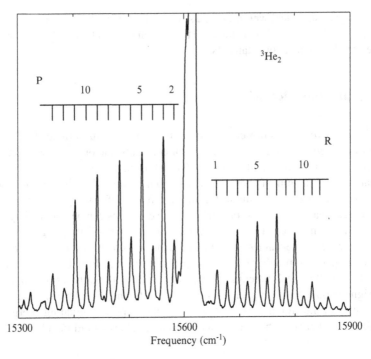

Fig. 6.4 Emission spectrum of $^3\text{He}_2$ near 640 nm

difference between the $J = 2$ and $J = 4$ levels of the lower potential for $v = 0$. The relevant equation is number (6.18). Using just the linear term and specializing to $v = 0$ yields

$$F_0(J_{upper}) - F_0(J_{lower}) = B_0(J_u(J_u + 1) - J_l(J_l + 1)).$$

If one does not have the quoted line positions and are reading values from a graph, it makes statistical sense to use two well-separated lines. Choosing $R(10)$ and $P(12)$, the energy difference is about 330 cm^{-1} which yields a value for $B_0 = 7.17$ cm^{-1}. This compares favorably to the quoted value of 7.306 cm^{-1}. One could do several pairs and the exercise is easier if the line positions are tabulated, which they often are.

Problem 6.7

Choose two well-separated lines from the spectrum of $^3\text{He}_2$ to obtain the B_0 value for the lower potential and compare to the quoted value of 9.67 cm^{-1}. Do the same for the upper potential and compare to 9.58 cm^{-1}.

Appendix A
Some Atomic Constants

Quantity	Symbol	Value in SI (cgs) units[a]
Speed of light in vacuum	c	2.99792458×10^8 m/s (10^{10} cm/s)
Elementary charge	e	$1.6021765 \times 10^{-19}$ C (4.803242×10^{-10} esu)
Planck's constant	h	6.626069×10^{-34} J s ($\times 10^{-27}$ erg s)
	$\hbar$	$1.0545716 \times 10^{-34}$ J s ($\times 10^{-27}$ erg s)
Electron rest mass	m_e	9.109382×10^{-31} kg ($\times 10^{-28}$ g)
Boltzmann constant	k_B	1.380650×10^{-23} J/K ($\times 10^{-16}$ erg/K)
	k_B/hc	$(0.6950356$ cm^{-1} K$^{-1})$
Rydberg constant	R_∞	$1.09737315685 \times 10^7$ m^{-1} ($\times 10^5$ cm^{-1})
	$R_\infty hc$	2.179872×10^{-18} J $= 13.605691$ eV
Fine-structure constant	α^{-1}	137.0359997
Bohr radius	a_0	$0.529177208 \times 10^{-10}$ m ($\times 10^{-8}$ cm)
Atomic mass unit	$1 \, u = m_u$	$1.6605388 \times 10^{-27}$ kg ($\times 10^{-24}$ g)
Proton rest mass	m_p	$1.6726216 \times 10^{-27}$ kg ($\times 10^{-24}$ gm)
	m_p/m_e	1836.152672
Electron g factor	g_e	-2.002319304362
Bohr magneton	μ_B	9.274009×10^{-24} J T^{-1}
	μ_B/hc	$(4.668645 \times 10^{-5}$ cm^{-1} gauss$^{-1})$
Nuclear magneton	μ_N	$5.0507832 \times 10^{-27}$ J T^{-1}

[a]P.J. Mohr, B.N. Taylor and D.B. Newell, Rev. Mod. Phys. **80**, 633 (2008)

R.L. Brooks, *The Fundamentals of Atomic and Molecular Physics*, Undergraduate
Lecture Notes in Physics, DOI 10.1007/978-1-4614-6678-9,
© Springer Science+Business Media New York 2013

Appendix B
Polynomials and Spherical Harmonics

The associated Laguerre polynomials are defined as

$$L_\lambda^\mu(x) = \frac{1}{\lambda!} x^{-\mu} e^x \frac{d^\lambda}{dx^\lambda} \left(x^{\lambda+\mu} e^{-x} \right)$$

The Legendre polynomials are defined as

$$P_\ell = \frac{1}{2^\ell \ell!} \frac{d^\ell}{dx^\ell} \left(x^2 - 1 \right)^\ell; \qquad P_\ell(1) = 1 \quad \text{for all} \quad \ell$$

The spherical harmonics are

$$Y_\ell^m(\theta, \phi) = (-1)^m e^{im\phi} \left[\frac{(2\ell+1)(\ell-m)!}{4\pi(\ell+m)!} \right]^{1/2} \sin^m \theta \frac{d^m}{dx^m} P_\ell(x) \quad \text{for} \quad x = \cos\theta$$

Orthonormality and completeness are given by

$$\int_0^{2\pi} \int_0^\pi Y_\ell^m(\theta, \phi) Y_{\ell'}^{*m}(\theta, \phi) \sin\theta \, d\theta \, d\phi = \delta_{\ell\ell'} \delta_{mm'} \qquad (B.1)$$

$$\sum_{\ell=0}^\infty \sum_{m=-\ell}^\ell Y_\ell^{*m}(\theta', \phi') Y_\ell^m(\theta, \phi) = \delta(\phi - \phi')\delta(\cos\theta - \cos\theta') \qquad (B.2)$$

R.L. Brooks, *The Fundamentals of Atomic and Molecular Physics*, Undergraduate Lecture Notes in Physics, DOI 10.1007/978-1-4614-6678-9,
© Springer Science+Business Media New York 2013

$$Y_\ell^{*m}(\theta, \phi) = (-1)^m Y_\ell^{-m}(\theta, \phi) \tag{B.3}$$

$$Y_\ell^m(\pi - \theta, \phi + \pi) = (-1)^\ell Y_\ell^m(\theta, \phi) \quad \text{Inversion: } \vec{r} \to -\vec{r} \tag{B.4}$$

$$Y_\ell^0(\theta, \phi) = \left(\frac{2\ell + 1}{4\pi}\right)^{1/2} P_\ell(\cos \theta) \quad \text{No } \phi \text{ dependence} \tag{B.5}$$

$$Y_\ell^m(0, \phi) = Y_\ell^0(0)\delta_{m0} = \sqrt{\frac{2\ell + 1}{4\pi}}\delta_{m0} \tag{B.6}$$

The expansion of $\frac{1}{r_{12}}$ occurs often in this text, and its derivation will be given here:

$$\frac{1}{|\vec{r}_1 - \vec{r}_2|} = \sum_{\ell=0}^{\infty} \sum_{m=-\ell}^{\ell} \frac{4\pi}{(2\ell + 1)} \frac{r_<^\ell}{r_>^{\ell+1}} Y_\ell^{*m}(\theta_2, \phi_2) Y_\ell^m(\theta_1, \phi_1) \tag{B.7}$$

$\vec{r}_1$ and $\vec{r}_2$ are arbitrary vectors having the usual spherical coordinate angles θ_1, ϕ_1 and θ_2, ϕ_2, respectively. Let γ be the angle between these vectors, $\hat{r}_1 \cdot \hat{r}_2 = \cos\gamma$, such that if the z-axis of a coordinate system were aligned with either $\vec{r}_1$ or $\vec{r}_2$, γ would play the role of θ for that coordinate frame. In that coordinate frame the role of ϕ is played by ω. The first step is to recall that

$$\nabla^2(|\vec{r}_1 - \vec{r}_2|^{-1}) = 0 \quad \text{except at } \vec{r}_1 = \vec{r}_2$$

If $\vec{r}_2$ is chosen to lie along the z-axis, there is azimuthal symmetry with the solution

$$\frac{1}{|\vec{r}_1 - \vec{r}_2|} = \sum_{\ell=0}^{\infty}[A_\ell r^\ell + B_\ell r^{-(\ell+1)}] P_\ell(\cos\theta)$$

This is the general solution to Laplace's equation in spherical coordinates with azimuthal symmetry. Since this solution is valid everywhere (except at $\vec{r}_1 = \vec{r}_2$), it must be valid for $\vec{r}_1$ on the z-axis. Then

$$\text{RHS} = \sum_{\ell=0}^{\infty}[A_\ell r^\ell + B_\ell r^{-(\ell+1)}]$$

$$\text{LHS} = \frac{1}{(r_1 - r_2)} = \frac{1}{r_1}(1 - r_2/r_1)^{-1} \quad r_1 > r_2$$

$$= \frac{1}{r_1}\left[1 + \frac{r_2}{r_1} + \left(\frac{r_2}{r_1}\right)^2 + \left(\frac{r_2}{r_1}\right)^3 + \cdots\right]$$

$$= \frac{1}{r_1}\sum_{\ell=0}^{\infty}\left(\frac{r_2}{r_1}\right)^\ell = \sum_{\ell=0}^{\infty}\frac{r_2^\ell}{r_1^{\ell+1}}$$

which holds whenever $r_1 > r_2$. Whenever $r_2 > r_1$, one obtains

$$\text{LHS} = \sum_{\ell=0}^{\infty} \frac{r_1^{\ell}}{r_2^{\ell+1}}$$

These two possibilities can be combined into the single expression:

$$\text{LHS} = \sum_{\ell=0}^{\infty} \frac{r_<^{\ell}}{r_>^{\ell+1}}$$

where $r_<(r_>)$ is the lesser (greater) of r_1 and r_2. This is compatible with the right-hand side. For example, if $r_1 > r_2$ $B_\ell = r_2^\ell$ and $A_\ell = 0$ while if $r_2 > r_1$ $A_\ell = 1/r_2^{(\ell+1)}$ and $B_\ell = 0$. When r_1 is not along the z-axis, the solution would look like

$$\frac{1}{|\vec{r}_1 - \vec{r}_2|} = \sum_{\ell=0}^{\infty} \frac{r_<^{\ell}}{r_>^{\ell+1}} P_\ell(\cos\theta)$$

Finally if r_2 had not been along the z-axis, θ would have been γ yielding

$$\frac{1}{|\vec{r}_1 - \vec{r}_2|} = \sum_{\ell=0}^{\infty} \frac{r_<^{\ell}}{r_>^{\ell+1}} P_\ell(\cos\gamma). \tag{B.8}$$

It remains to show that $P_\ell(\cos\gamma)$ can be expanded in spherical harmonics. The expression is referred to as the spherical harmonic addition theorem:

$$P_\ell(\cos\gamma) = \left(\frac{4\pi}{2\ell+1}\right) \sum_{m=-\ell}^{\ell} Y_\ell^{*m}(\theta_2, \phi_2) Y_\ell^m(\theta_1, \phi_1) \tag{B.9}$$

First consider a function $g(\theta, \phi)$ which will at first be identified with the spherical harmonic having coordinates θ_1, ϕ_1. It will then be expanded in spherical harmonics using γ, ω coordinates. It's value at $\gamma = 0$ will prove to be important, but at that value for γ, it becomes equal to the spherical harmonic having coordinates θ_2, ϕ_2. Let's see how this unfolds:

$$g(\theta_1, \phi_1) \equiv Y_\ell^m(\theta_1, \phi_1) \tag{B.10}$$

$$= \sum_{m'=-\ell}^{\ell} a_{\ell m'} Y_\ell^{m'}(\gamma, \omega) \tag{B.11}$$

No summation over ℓ is needed as the spherical harmonics do not change ℓ value under a coordinate rotation:

$$g(\theta_1, \phi_1)|_{\gamma=0} = \sum_{m'=-\ell}^{\ell} a_{\ell m'} \left[\frac{(2\ell+1)}{4\pi}\right]^{1/2} \delta_{m'0} = a_{\ell 0} \left[\frac{(2\ell+1)}{4\pi}\right]^{1/2} \tag{B.12}$$

This follows from property (B.6) of spherical harmonics. Using (B.11), one can see that

$$\int g(\theta_1, \phi_1) \, Y_\ell^{*0}(\gamma, \omega) \, d\Omega_{\gamma, \omega} = a_{\ell 0}$$

But from (B.10), this means that

$$\int Y_\ell^m(\theta_1, \phi_1) \, Y_\ell^{*0}(\gamma, \omega) \, d\Omega_{\gamma, \omega} = a_{\ell 0} \tag{B.13}$$

It is now possible to expand $P_\ell(\cos \gamma)$ itself in spherical harmonics:

$$P_\ell(\cos \gamma) = \sum_{m'=-\ell}^{\ell} b_{\ell m'} \, Y_\ell^{m'}(\theta_1, \phi_1) \tag{B.14}$$

If one now multiplies both sides by Y_ℓ^{*m} and integrates over all space,

$$\int P_\ell(\cos \gamma) Y_\ell^{*m} \, d\Omega = \sum_{m'=-\ell}^{\ell} b_{\ell m'} \, \delta_{mm'} = b_{\ell m} \tag{B.15}$$

From Equation (B.5), it follows that

$$P_\ell(\cos \gamma) = \left[\frac{4\pi}{(2\ell+1)} \right]^{1/2} Y_\ell^0(\gamma, \omega)$$

though in this expression ω is irrelevant. Inserting this into (B.15) yields

$$\left[\frac{4\pi}{(2\ell+1)} \right]^{1/2} \int Y_\ell^0(\gamma, \omega) Y_\ell^{*m}(\theta_1, \phi_1) \, d\Omega = b_{\ell m}$$

But from (B.13), it follows that

$$b_{\ell m}^* = a_{\ell 0} \left[\frac{4\pi}{(2\ell+1)} \right]^{1/2}$$

Substituting the right-hand side of the above from (B.12) yields

$$b_{\ell m}^* = \frac{4\pi}{(2\ell+1)} \, g(\theta_1, \phi_1) |_{\gamma=0}$$

But as stated in the introduction of this derivation at $\gamma = 0$, one can write

$$g(\theta_1, \phi_1) |_{\gamma=0} = Y_\ell^m(\theta_2, \phi_2)$$

from which it follows that

$$b_{\ell m}^* = \frac{4\pi}{(2\ell+1)} Y_\ell^m(\theta_2, \phi_2).$$

Taking the complex conjugate and putting back into Eq. (B.14) yields the result

$$P_\ell(\cos\gamma) = \left(\frac{4\pi}{2\ell+1}\right)\sum_{m=-\ell}^{\ell} Y_\ell^{*m}(\theta_2,\phi_2)Y_\ell^m(\theta_1,\phi_1)$$

where the dummy index m' has been replaced by m everywhere. This completes the derivation of Equation (B.7).

Sometimes one sees spherical harmonics redefined to emphasize their relations to the Cartesian coordinates x, y, and z. Define

$$C_\ell^m \equiv \left(\frac{2\ell+1}{4\pi}\right)^{1/2} Y_\ell^m(\theta,\phi)$$

(eliminates some annoying constants.) Note that $C_1^0 = \cos\theta = Z/r$:

$$\frac{C_1^{-1} - C_1^1}{\sqrt{2}} = \sin\theta\cos\phi = x/r$$

$$\frac{C_1^{-1} + C_1^1}{-\sqrt{2}i} = \sin\theta\sin\phi = y/r$$

These linear combinations have the spatial symmetries of x, y, and z. Wave functions using these combinations are labeled p_x, p_y, and p_z (p because $\ell = 1$).

Similarly for $\ell = 2$, one may write

$$C_2^0 = \frac{3}{2}\cos^2\theta - \frac{1}{2} = \frac{1}{r^2}\left(z^2 - \frac{x^2+y^2}{2}\right)$$

$$\frac{C_2^{-1} - C_2^1}{\sqrt{2}} = \sqrt{3}\sin\theta\cos\theta\cos\phi = \sqrt{3}\frac{xz}{r^2}$$

$$\frac{C_2^{-1} + C_2^1}{-\sqrt{2}i} = \sqrt{3}\sin\theta\cos\theta\sin\phi = \sqrt{3}\frac{yz}{r^2}$$

$$\frac{C_2^2 = C_2^{-2}}{\sqrt{2}i} = \frac{\sqrt{3}}{2}\sin^2\theta\sin 2\phi = \sqrt{3}\frac{xy}{r^2}$$

$$\frac{C_2^{-2} + C_2^2}{\sqrt{2}} = \frac{\sqrt{3}}{2}\sin^2\theta\cos 2\phi = \frac{\sqrt{3}}{2}\frac{x^2-y^2}{r^2}$$

These linear combinations are sometimes labeled d_{xy}, $d_{x^2-y^2}$, etc. This labeling is not usually carried beyond $\ell = 2$. Such wave functions are often used for molecular orbital theory.

Appendix C
Some Tensor Background

A vector may be defined as any object which transforms like a coordinate point

$$A'_i = \lambda_{ij} A_j$$

A coordinate point transforms by coordinate rotation by

$$x'_i = \lambda_{ij} x_j$$

where $\lambda_{ij} \equiv \cos(x'_i, x'_j)$.

In n-dimensional space, an mth rank tensor is an object which transforms under coordinate rotations as

$$T'_{abcd...} = \lambda_{ai}\lambda_{bj}\lambda_{ck}\lambda_{d\ell} \ldots T_{ijkl...}$$

It has n^m components. Such a Cartesian tensor has a rank given by the number of indices. In three dimensions, an ℓth-rank tensor has 3^ℓ components.

A symmetric tensor is invariant to the interchange of any two indices. For an ℓth-rank tensor, this reduces the number of components from 3^ℓ to $(\ell+1)(\ell+2)/2$. (Can you show this?) For example, a 4th rank tensor is reduced from 81 to 15 components.

Now a second rank tensor is traceless whenever

$$\delta_{ij}T_{ij} = 0 \quad \text{or} \quad T_{11} + T_{22} + T_{33} = 0$$

The generalization of this is that

$$\delta_{mn}T_{ijk...\ell} = 0$$

where m and n are *any* two indices. Such a tensor is said to be irreducible and has only $(2\ell + 1)$ independent components. So a 4th rank tensor which started with 81 components would have only 9.

Most tensors which describe physical phenomena are symmetric, and by being clever, one can usually make them irreducible.

R.L. Brooks, *The Fundamentals of Atomic and Molecular Physics*, Undergraduate
Lecture Notes in Physics, DOI 10.1007/978-1-4614-6678-9,
© Springer Science+Business Media New York 2013

Consider, for example, an electrostatic multipole moment. You may recall that the quadrupole moment is defined as

$$Q_{ij} = \frac{1}{2} \int \rho(\vec{r}') \left(3x_j' x_i' - r'^2 \delta_{ij} \right) d\tau'$$

The $2^{\ell\text{th}}$ pole moment is defined as

$$Q_{ijk\ldots\ell} \equiv \frac{(-1)^\ell}{\ell!} \int \rho(\vec{r}') r'^{(2\ell+1)} \nabla_i' \nabla_j' \nabla_k' \cdots \nabla_\ell' \left(\frac{1}{r'} \right) d\tau'.$$

Such a moment satisfies $\delta_{mn} Q_{ijk\ldots\ell} = 0$ and is symmetric.

Recall that for Y_ℓ^m m ranges from $-\ell$ to ℓ and takes on $(2\ell + 1)$ values. In this way Y_ℓ^m can be used as a *basis* for irreducible tensors or spherical tensors. The spherical tensor analog of $Q_{ijk\ldots\ell}$ is

$$q_{lm} \equiv \int Y_\ell^{*m}(\theta', \phi') r'^\ell \rho(\vec{r}') d\tau'$$

Appendix D
Magnetic Dipole Interaction Energy

Recall that the definition of the magnetic dipole, $\vec{\mu}$, of a current distribution is

$$\vec{\mu} \equiv \frac{1}{2c} \int \vec{r'} \times \vec{J}(\vec{r}) \, d\tau'$$

But

$$\vec{J} = Nq\vec{v} = N\frac{q}{m}\vec{p}$$

where N is the number of particles (of mass m and charge q) per unit volume and $\vec{p}$ is the momentum. So

$$\vec{\mu} = \frac{Nq}{2cm} \int (\vec{r'} \times \vec{p'}) \, d\tau'$$

If there is but one particle in a volume V with charge $q = -e$ whose angular momentum is a constant of the motion, the dipole moment may be written as

$$\vec{\mu} = -\frac{e\vec{\ell}}{2cmV} \int d\tau' = -\frac{e\vec{\ell}}{2mc} \tag{D.1}$$

The interaction energy (potential energy) of a magnetic dipole moment in an external magnetic field is what is desired. (The analogous result for an electric dipole in an external electric field is $-\vec{p}\cdot\vec{E}$.) Expand the magnetic field about some suitable origin:

$$B_i(\vec{r}) = B_i(0) + \vec{r} \cdot \vec{\nabla} B_i(0) + \cdots \tag{D.2}$$

Now the force on a current distribution in an external field is

$$\vec{F} = \frac{1}{c} \int \vec{J}(\vec{r'}) \times \vec{B}(\vec{r'}) \, d\tau' \tag{D.3}$$

(This is just an extension of the Lorentz law, $\vec{F} = (q/c)\vec{v} \times \vec{B}$.) Putting (D.2) into (D.3) gives

$$\vec{F} = \frac{-1}{c}\vec{B}(0) \times \int \vec{J}(\vec{r'}) \, d\tau' + \frac{1}{c} \int \vec{J}(\vec{r'}) \times [(\vec{r'} \cdot \vec{\nabla})\vec{B}(0)] \, d\tau' + \cdots$$

R.L. Brooks, *The Fundamentals of Atomic and Molecular Physics*, Undergraduate Lecture Notes in Physics, DOI 10.1007/978-1-4614-6678-9,
© Springer Science+Business Media New York 2013

The first term is zero for steady-state localized currents. Next note that

$$\vec{J}(\vec{r'}) \times [(\vec{r'} \cdot \vec{\nabla})\vec{B}] = \vec{J}(\vec{r'}) \times \vec{\nabla}(\vec{r'} \cdot \vec{B})$$

This follows by the vector identity

$$\vec{\nabla}(\vec{r'} \cdot \vec{B}) = \vec{r'} \times (\vec{\nabla} \times \vec{B}) + \vec{B} \times (\vec{\nabla} \times \vec{r'}) + (\vec{r'} \cdot \vec{\nabla})\vec{B} + (\vec{B} \cdot \vec{\nabla})\vec{r'}$$

However, $\vec{\nabla} \times \vec{B} = 0$ and ∇ do not operate on primed variables, so only the third term on the RHS is nonzero. Next note that

$$\vec{\nabla} \times (\vec{r'} \cdot \vec{B})\vec{J} = (\vec{r'} \cdot \vec{B})\vec{\nabla} \times \vec{J}(\vec{r'}) + \vec{\nabla}(\vec{r'} \cdot \vec{B}) \times \vec{J}(\vec{r'})$$

This is a vector identity and the first term of the RHS is zero because ∇ does not operate on $\vec{J}(\vec{r'})$. So

$$\vec{F} = -\frac{1}{c}\vec{\nabla} \times \int \vec{J}(\vec{r'})(\vec{r'} \cdot \vec{B}(0)) \, d\tau' \tag{D.4}$$

Now use the identity

$$\vec{B} \times (\vec{r'} \times \vec{J'}) = \vec{r'}(\vec{B} \cdot \vec{J'}) - \vec{J'}(\vec{r'} \cdot \vec{B})$$

to express the integral as

$$\int \vec{J}(\vec{r'})(\vec{r'} \cdot \vec{B}) \, d\tau' = \int \vec{r'}(\vec{B} \cdot \vec{J'}) \, d\tau' - \vec{B} \times \int (\vec{r'} \times \vec{J'}) \, d\tau' \tag{D.5}$$

On the LHS, there is

$$B_i \int J'_j x'_i \, d\tau' = B_i \int \left[\nabla'_\ell (x'_j J'_\ell) \right] x'_i \, d\tau'$$

(This is easy to get by working on the right to obtain the left.) Now integrate the RHS by parts:

$$= -B_i \int x'_j J'_\ell \nabla'_\ell x'_i \, d\tau'$$

$$= -B_i \int x'_j J'_i \, d\tau'$$

$$= -\int \vec{r'}(\vec{B} \cdot \vec{J'}) \, d\tau'$$

So the first term on the RHS of (D.5) is the negative of the LHS. (D.5) becomes

$$\int \vec{J}(\vec{r'})(\vec{r'} \cdot \vec{B}) \, d\tau' = -\frac{1}{2}\vec{B} \times \int (\vec{r'} \times \vec{J'}) \, d\tau'$$

Putting this into (D.4) gives

$$\vec{F} = \vec{\nabla} \times \left[\vec{B} \times \frac{1}{2c} \int (\vec{r'} \times \vec{J'})\, d\tau' \right]$$

$$\text{or} \quad \vec{F} = \vec{\nabla} \times (\vec{B} \times \vec{\mu}) \tag{D.6}$$

Now use the vector identity

$$\vec{\nabla} \times (\vec{A} \times \vec{B}) = \vec{A}(\vec{\nabla} \cdot \vec{B}) - \vec{B}(\vec{\nabla} \cdot \vec{A}) + (\vec{B} \cdot \vec{\nabla})\vec{A} - (\vec{A} \cdot \vec{\nabla})\vec{B}$$

$$\text{and} \quad \vec{F} = (\vec{\mu} \cdot \vec{\nabla})\vec{B} = \vec{\nabla}(\vec{\mu} \cdot \vec{B}) \tag{D.7}$$

remembering that $\vec{\mu}$ is a constant vector and that div $\vec{B} = $ curl $\vec{B} = 0$.
So if $\vec{F} = -\nabla W$ where W is the potential energy, it follows that

$$W = -\vec{\mu} \cdot \vec{B} \tag{D.8}$$

Index

R.L. Brooks, *The Fundamentals of Atomic and Molecular Physics*, Undergraduate
Lecture Notes in Physics, DOI 10.1007/978-1-4614-6678-9,
© Springer Science+Business Media New York 2013